HVAC Project Management Simplified

A complete field guide for engineers, managers, and contractors to deliver projects on time, on budget, and with quality

Dr. RVSS Prasad

gurucool
PUBLISHING

COPYRIGHT

Published by:
Gurucool Publishing
#102, Sai Krupa Nilayam,
Nagendra Nagar, Habsiguda, Hyderabad - 500007
Ph: 7702566640
E-mail: info@gurucoolpublishing.com

ISBN: 978-93-49999-67-1

First Edition - 2025

Book Cover by Varsha Singh
Illustrations by Varsha Singh

DEDICATION

To My Parents
Late: Sri R.V Seshaiah &
Smt. R.A. Lakshmi
and
To My mentor
Prof Dr. A. RAMACHANDRA ARYASRI

ACKNOWLEDGEMENTS

This book is the culmination of years of learning, collaboration, and inspiration from many remarkable individuals who have shaped my personal and professional journey.

My deepest gratitude goes to my family — my mother Lakshmi, Sobha Rani, Dr. Cody Stevens, Sreelekha, Sreeja, and Dr. Ram — for their unwavering love, strength, and encouragement. Your belief in me has been the foundation upon which this work was built.

I extend my sincere thanks to my friends and partners at Pavani Engineers — V. Subramanian and B.V.N. Satyanarayana Kumar — and to the entire team for their trust, teamwork, and constant motivation that kept this vision alive.

I am profoundly grateful to my industry mentors — V. Sridhar, M. Gopi Krishna, K. Ramesh Babu, Prashant Raidu, Senthil Kumar, and Captain Prasad — as well as my dear friends from VRSEC, Vijayawada. Your wisdom, guidance, and faith in my abilities have been invaluable throughout this journey.

A special note of thanks to Mr. Chandrasekar Narayanan, Consultant & Certified Energy Auditor, Genex Consultants, for graciously writing the foreword and for his steadfast encouragement and support.

My heartfelt appreciation to Mr. Shivamohan, whose motivation and belief in my work served as a powerful catalyst for bringing this book to life.

I also wish to acknowledge the professional communities of ISHRAE, ASHRAE, and PMIPCC, along with Professor A. R. Aryasri, Professor G.V.K. Acharyulu, and many esteemed friends and colleagues whose insights and camaraderie have enriched this endeavour.

To everyone who believed in me and walked alongside me on this path — my sincerest thanks and enduring gratitude.

Lastly, I acknowledge the support of ChatGPT for assisting in structuring and refining sections of this manuscript, helping make the book more cohesive and reader friendly.

FOREWARD

The author of this book, Dr RVSS Prasad, is a versatile professional in the field of Project Management and Execution who has a unique identity as an expert in this field and exposure to multiple verticals. He is personally known to me for his consistent involvement in different HVACR projects of varying magnitude, timelines and complications, and he has worked closely in many of the projects under my consultancy as well as with other corporations. He has ensured his vast experience and exposure is brought forward in this book.

Readers of this book can expect to understand details from a simple statement on Project Management, why needed and subsequently understand the detailed role of the Project Manager, who is the architect of Project Planning and Execution. His detailed explanation of the Scope of planning and Work Breakdown Structure will be greatly helpful in ensuring Project Quality Delivery.

A well-written chapter on Testing and Commissioning is an asset to every Project Owner, as this is least emphasised in many projects, small or big and invariably ignored due to closure needs. The chapter on this topic brings out intricate details that are required to be followed.

His detailing on Project Closure and Administrative Closure is a must-read and implement. His statement that Project closing leaves the final and lasting impression on the Clients are statements which have come out of the author's personal experience and should receive the attention it deserves.

Overall, a well-documented book on HVAC project management which gives enough reasons to ensure Owners / PMC / Architects / Developers / Consultants / Contractors and every team member contributing to the Project have utility value to gain, which shall ensure quality outcomes obtained for project needs.

A must-read, understand and adopt for all Building Services Professionals

CHANDRASEKAR NARAYANAN SRIKANTAN
Consultant & Certified Energy Auditor, Genex Consultants, Hyderabad
Certified Engineering professional (Consultant)

PREFACE

"HVAC Projects Management Simplified"

Why this book?

In most engineering colleges and training programs, HVAC systems are taught as technical subjects — Thermodynamics, heat load calculations, psychometrics, duct design, or piping hydraulics. But the real world of HVAC projects is very different. Once a project moves from paper to site, engineers face a completely new set of challenges: *delays in approvals, missing site readiness, clashes with other trades, customs delays for imported chillers, subcontractor disputes, safety compliance, and client escalations.*

Too often, young engineers are thrown into projects without a roadmap. They learn the hard way by trial and error, with mistakes that cost companies money, time, and reputation. Even senior engineers who have mastered design or installation often find project management elusive: they can solve technical problems but struggle with contracts, costs, schedules, or coordination.

This book was born out of that gap. It is not **another textbook on HVAC design**. It is a **practical field guide** to **Project Management for HVAC engineers**.

What is this book all about?

- A practical, phase-by-phase guide to HVAC project management across initiating, planning, executing, erection, testing, commissioning, handover, and close-out, grounded in real site conditions and stakeholder dynamics.

- Tools, checklists, templates, and best practices to plan work, manage scope, control quality, align procurement, and keep schedules and cash flow on track.

- Core competencies for project managers—leadership, vendor and subcontractor management, contract awareness, reporting, change control, and communication under pressure

Who is this book for?

- **Fresh graduates** — who know theory but want to understand *how projects really work*.
- **Site engineers & supervisors** — who need a step-by-step guide for drawings, approvals, submittals, installation, and coordination.
- **PMs, Senior managers & contractors** — Experienced professionals looking to standardise lessons learned, uplift subcontractor performance, and mentor teams with proven frameworks
- **PMCs** responsible for integrating MEP trades, enforcing documentation rigour, and driving timely, claim-resistant delivery
- **Consultants**— Consultants aiming for smoother design-to-site transitions, clearer submittal pathways, and verifiable commissioning outcomes.
- **Client representatives** who need transparent baselines, milestone visibility, risk flags, and defensible decisions across contractors

In short, this book is for **any HVAC professional who wants to grow from "doing the work" to "managing the work" — efficiently, ethically, and profitably.**

Why this book?

- Converts scattered tasks into a repeatable operating system—kick-off to close-out—with ready-to-use formats for reporting, commissioning, vendor evaluation, and handover.
- Anticipates common pitfalls—scope creep, interface clashes, late procurement, and documentation gaps—and shows how to prevent them before they cost time and money.
- Aligns modern practices—BIM, digital tracking, EVM thinking, and energy-efficiency compliance—with day-to-day site realities for tangible results.

What this book is not?

- It is not a technical treatise; detailed heat load calculations, equipment selection methodologies, clean rooms and equipment troubleshooting are intentionally outside the scope.
- It does not replace OEM manuals, design codes, or commissioning standards; rather, it shows how to plan, coordinate, verify, and document their application in projects.
- It is not a claims textbook; instead, it equips readers with the contract awareness and documentation discipline that reduces disputes in the first place
- The templates given are for understanding purposes only. Every project needs to have its own checklists and templates based on the contract agreements and organisation culture.

How to use this book?

- This book is structured to guide you through the **entire lifecycle of an HVAC project**, from the first tender review to project closeout and lessons learned.
- Read end-to-end to build a mental model of the full lifecycle, then return to the checklists and templates to run live projects with confidence.
- Treat each phase chapter as a go-to playbook—initiation for kick-off and scope clarity, planning for submittals and scheduling, execution for site controls, commissioning for verifications, and handover for a clean close.
- Leverage the scenarios and lessons to coach teams, align vendors, and hold productive coordination meetings that convert plans into performance

What is covered?

- **Part I – Foundations**: Builds the mindset — why project management matters, the role of an HVAC project manager, and the lifecycle framework.
- **Part II – Planning Core**: The backbone of success — WBS, drawings, technical submittals, method statements, scheduling, budgeting, procurement, quality, risk, and integration.
- **Part III – Execution & Control**: The field manual — covering project execution, commissioning, and monitoring/controlling. Packed with Do's & Don'ts, checklists, and trackers.

- **Part IV – Completion & Handover**: Ensures projects close well — lessons learned, contracting essentials, CCPM, and the final wrap-up.
- **Part V – Annexures**: Templates, trackers, checklists, and sample documents for immediate use.

Each chapter includes **Golden Rules, case studies, summary sheets, and annexures** to make it practical. You can either read it sequentially, as a textbook, or dip into specific sections when working on site.

Why now?

The HVAC industry is evolving faster than ever. With green building standards, digital tools (BIM, IoT, analytics), stricter safety norms, and tighter margins, *project management has become as important as technical design*. Engineers who master both will not only survive but thrive in this competitive space.

Final Word!

This book is written for every HVAC professional who has stood in a noisy plant room, a dusty site, or a tense project meeting and thought: *"There has to be a better way to manage this."*

If the essence of HVAC project success is disciplined coordination, this handbook is the field companion that makes discipline practical—clear enough to start tomorrow, robust enough to scale across projects, and adaptable enough to meet the demands of modern delivery.

- Dr RVSS Prasad

Content

List of Tables

List of figures

Chapter 1: Need for HVAC Project Management

"Failing to plan is planning to fail — especially in HVAC projects, where every delay or error multiplies across trades and timelines."

Learning Objectives

By the end of this chapter, the reader will be able to:

1. Understand why project management is critical in HVAC projects.

2. Recognise common pitfalls in HVAC projects without structured project management.

3. Appreciate the benefits of applying project management principles to HVAC projects.

4. Relate project management concepts to real-world HVAC industry examples.

1.1 Introduction

"In HVAC, success is not defined by the drawings we create, but by how well the system performs to its design intent after handover."

The Heating, Ventilation, Air Conditioning, and Refrigeration (HVAC&R) industry in India has a history spanning over 80 years. From niche luxury to an indispensable necessity, HVAC has grown into the backbone of modern infrastructure —

hospitals, IT campuses, pharma plants, and green buildings. Since the early 21st century, the industry's growth has been exponential, fueled by the IT boom, rapid urbanisation, and rising demand for controlled indoor environments.

1.2 Current Trends and Global Outlook

Leaders in the HVAC sector must stay aware of evolving standards (latest ISHRAE, ASHRAE, and international norms for sustainability and energy efficiency) to be credible and competitive. The Indian HVAC landscape is seeing convergence with global practices through regulation and client expectations, especially in green buildings, IEQ benchmarks, and advanced climate control systems.

1.3 Scale of Projects and Complexity

The scale of projects has expanded dramatically, while delivery timelines have compressed — shifting from months or even years to just a few weeks in some cases. HVAC projects are complex, multidisciplinary, and time sensitive. Unlike standalone technical jobs, HVAC projects interact with civil, electrical, and finishing works. For example, delays in duct approval may hold back ceiling closure; late chiller delivery can delay the entire building handover.

Anyone who has worked in HVAC knows the reality: projects often **struggle with rework, delays in approvals, misaligned drawings, cash flow disputes, or last-minute commissioning firefighting. These are not technical failures — they are failures of project management.**

In such an environment, project management is no longer optional; it is the foundation of success. It ensures that all stakeholders, from clients and consultants to contractors and suppliers, work cohesively towards a shared goal. Without it, even the most well-designed system can falter in execution, leading to delays, budget overruns, and dissatisfied end-users.

This book is written to address exactly that gap.
It is for:

- Young engineers are seeking to grow into project managers.
- Experienced professionals who want a refresher and practical tools.
- Contractors and consultants who juggle multiple stakeholders daily.

Through Indian case studies, practical checklists, and ready-to-use templates, this book brings the **Project Management Lifecycle** alive — guiding a project from initiation through planning, execution, monitoring, and controlling, and finally, commissioning and handover. Readers will walk away not with theory, but with tools they can apply on site the very next day.

> **Golden Rule:** *"A project without structured management is like a building without foundations — it may stand temporarily, but it won't last."*

1.4 Digital Transformation and Modern Challenges

The HVAC industry is in the midst of rapid digitisation. BIM (Building Information Modelling), IoT-connected devices, remote monitoring, and AI-driven analytics now allow practitioners to design, plan, and operate systems with unprecedented efficiency and transparency. Automation, digital dashboards, and data-driven diagnostics have become essential for timely delivery and high-quality service.

- BIM-based clash detection (duct vs beam vs cable trays).

- IoT sensors for continuous commissioning.
- Data loggers used for validating indoor air quality.

Project managers must understand digital skills as well as the integration of technology across workflows. For example, robust document management systems, cloud-based communication, and digital scheduling/tracking tools (such as MS Project, Primavera, and BIM 360) are increasingly expected by clients and consultants worldwide, including in India.

1.5 What is Project Management?

Project Management can be understood as a **systematic approach to planning, organising, monitoring, and controlling resources** to achieve well-defined, time-bound objectives. Every project carries inherent risks, but what makes it unique is that it always has a definite start date and a definite end date, within which the project objectives must be achieved.

Project management *is not just a science of following methodologies and checklists to expect results*. It is also an art, managing people from different cultures, collaborating across multiple agencies, resolving conflicts, and motivating everyone to work towards the common outcome. The larger and more complex the project is, the greater the need for structured project management processes.

Globally, **the Project Management Institute (PMI)** has established widely recognised frameworks through the PMBOK® Guide (Project Management Body of Knowledge), which it updates periodically. PMI standards and practices are used extensively in IT, construction, and infrastructure projects across the world.

The definition of Project Management as per PMBOK (Project Management Body of Knowledge, PMI):

***"Project management is the application of knowledge, skills, tools, and techniques to project activities to meet the project requirements."* (PMBOK® Guide, 6th Edition, PMI)**

Table 13.1 Project Management Science Vs Art

Science of Project Management	Art of Project Management
Following structured frameworks (e.g., PMI's PMBOK).	Building trust and motivating teams.
Using tools like WBS, Gantt charts, and EVM.	Handling conflicts across trades.
Defining scope, schedule, and cost baselines.	Adapting to diverse cultures and site realities.
Monitoring KPIs and variances.	Inspiring people to stay focused under pressure.
Applying processes for risk, quality, and procurement.	Negotiating, influencing, and problem-solving.

It emphasizes that project management is achieved through the appropriate application and integration of the 49 processes (in PMBOK 6th Ed., or 33 processes in PMBOK 7th Ed.), which are grouped into five process groups **(Initiating, Planning, Executing, Monitoring & Controlling, Closing)** and span across the ten knowledge areas **(scope, schedule, cost, quality, resource, communications, risk, procurement, stakeholder, and integration management).**

In India, however, the HVAC industry has been a **late adopter** of structured project management. While large construction and IT projects are aligned with PMI standards, many HVAC organisations have only recently begun to adopt these frameworks — often in a customised way. Increasingly, leading companies are setting up **Project Management Offices (PMOs)** to bring consistency, standardise templates, and build reusable tools. Continuous improvement cycles ensure that lessons learned are captured and applied across projects.

Why it matters for HVAC: Unlike industries that adopted project management early, HVAC is catching up. This book aims to bridge that gap — by tailoring global best practices to the **unique challenges and realities of HVAC projects in India.**

1.6 Why Project Management in HVAC?

HVAC projects today are not just about air-conditioning buildings — they are about delivering precision, compliance, and comfort under high-stakes conditions.

The complexity has risen dramatically:

1. Supply Chain & Resources

- Long-lead imports, volatile prices, and customs delays disrupt material flow.
- Scarcity of skilled specialists in balancing, BMS integration, and site supervision works increases execution risk.

2. Site & Coordination Challenges

- Space constraints, overlapping work fronts, and clashes with civil/electrical works demand tighter sequencing.
- Multiple stakeholders — consultants, contractors, vendors — add layers of dependencies.

3. Regulatory & Safety Requirements

- Energy codes (ECBC), fire safety, and environmental clearances must all be complied with.
- Stringent HSE norms — confined space permits, working-at-height, and PPE enforcement — leave no room for shortcuts.

4. Client & Market Pressures

- Clients expect measurable Indoor Environmental Quality (IEQ), sustainability compliance, and high efficiency.
- Timelines are shrinking, tolerance for overruns is near zero, and reputational stakes are higher than ever.

Together, these factors mean that "business as usual" is no longer enough. HVAC projects demand structured project management frameworks that can integrate design, procurement, execution, and commissioning into one disciplined roadmap — ensuring delivery on time, within budget, and to the promised performance.

> **Golden Rule:** *"HVAC projects don't fail during commissioning; they fail months earlier in planning and coordination."*

1.7 Benefits of Structured HVAC Project Management

A disciplined project management approach transforms these challenges into opportunities:

- **Clear WBS & schedules** → early visibility of dependencies and bottlenecks.
- **Risk identification** → avoiding surprises (e.g., customs delays for imported chillers).
- **Quality control processes** → catching deviations early, preventing costly rework.
- **Structured communication** → fewer disputes, faster approvals, and smoother handovers.

- **Cash Flow Discipline** → Many HVAC PMs underestimate how critical this is.

1.8 Common Failures in HVAC Projects Without Project Management?

When HVAC projects run without structured project management, the results are almost predictable:

- **Material chaos** → Long-lead items not ordered in time, delaying execution.
- **Design–execution mismatches** → Consultants' drawings do not match site conditions, leading to rework.
- **Scope creep** → Uncontrolled changes inflate cost and delay delivery.
- **Subcontractor mismanagement** → Teams mobilised late, sites left idle.
- **Measurement & BOQ disputes** → Payments delayed for months.
- **Rushed commissioning** → Systems handed over with unresolved snags, leading to client complaints.

In short: ***Without project management, HVAC projects drift; with it, they deliver.***

Case Study: A Success Story: Project Management Lifecycle in Action

In 2022, a **pharmaceutical manufacturing facility in Pune** needed a complex HVAC system to maintain precise temperature, humidity, and particulate levels across multiple cleanrooms. The project had a strict nine-month deadline tied to regulatory approvals.

The project team applied a rigorous "**Project lifecycle approach**:

- Early stakeholder alignment.
- Detailed schedule with procurement and commissioning milestones.
- Weekly coordination meetings across all trades.
- Active procurement tracking for long-lead equipment.
- Weekly coordination meetings across all trades.
- Active procurement tracking for long-lead equipment.

Despite supply chain delays in air-handling units, proactive monitoring enabled quick alternate sourcing. Commissioning finished **three weeks ahead of schedule**, and the client passed regulatory audits with zero HVAC-related observations.

1.9 Conclusion:

HVAC projects are no longer strictly straightforward engineering jobs; they are multi-trade, time-sensitive, regulation-bound challenges that demand discipline and foresight. Without structured project management, even the most sophisticated designs are reduced to chaos on site — delays in duct approvals, idle equipment waiting for clearances, cash flow blockages, and frustrated clients.

With project management, the same challenges are transformed into opportunities: risks are identified early, dependencies are sequenced logically, costs and cash flows are tracked, and communication keeps all stakeholders aligned. The difference between rework and reliability, between disputes and trust, is almost always project management.

This chapter establishes the "why." The rest of this book will take you through the "how" — turning frameworks into daily practices, so that every project you manage is not just completed, but completed with confidence, efficiency, and reputation intact.

At a Glance: Need for HVAC Project Management

Why Project Management in HVAC?

- Projects are increasingly complex, with multiple trades, tight deadlines, and high client expectations.
- Without PM → rework, disputes, delays, and cost overruns.
- With PM → clarity, control, accountability, and successful handovers.

Key Benefits of Structured PM

- Clear WBS & schedules → visibility of dependencies.
- Early risk identification → avoid surprises (e.g., import delays).
- Quality processes → detect issues early, reduce rework.
- Structured communication → faster approvals, fewer disputes.
- Financial control → protect margins and ensure smooth cash flow.

Common Pitfalls Without PM

- Long-lead materials not ordered → idle sites.
- Drawings/site misalignments → costly rework.
- Scope creep → inflated costs and delays.
- Subcontractors mobilized late → project stalls.
- Rushed commissioning → unresolved snags, client complaints.

Golden Rules

- "A project without structured management is like a building without foundations — it won't last."
- "HVAC success is not built on design alone — it is built on disciplined project management from day one."

Takeaway: Project management is not optional in HVAC — it is the foundation that turns drawings into performance, and contracts into client confidence.

Chapter 2:
Role of HVAC Project Manager

"The best Project Managers are often silent — their projects speak through quality, timeliness, and trust."

Learning Objectives

By the end of this chapter, the reader will be able to:

1. Identify the **primary responsibilities** of an HVAC Project Manager in the Indian market.

2. Recognise the **technical and managerial skills** essential for project success.

3. Understand how to effectively **coordinate with multiple stakeholders** across the project lifecycle.

4. Integrate **safety, compliance, and ethics** into daily project management activities.

5. Evaluate **effective vs. ineffective project manager behaviours**

2.1 Role of HVAC Project Manager

In HVAC projects, the Project Manager is the central figure who ensures that design intent, procurement, installation, and commissioning are executed seamlessly. In the Indian context, this **role demands more than technical knowledge; it**

requires commercial knowledge, leadership, adaptability, and an ability to manage multiple stakeholder complexities. A successful HVAC PM blends technical expertise with strong communication, risk management, and commercial acumen, all while ensuring quality standards.

Project Manager self-assessment tools are available through ASHRAE. ISHRAE and PMI chapters (e.g., Core PM Competency Self-Check, ISHRAE Project Manager Leadership Framework), etc.

Table 2. 1 Skills vs Standards

Standard/Framework	Technical Skills	PM Skills	Regulatory/ Compliance
ASHRAE	✓		✓
ISHRAE	✓		✓
PMI (PMBOK)		✓	

All project managers are encouraged to familiarise themselves with core modules in BIM, scheduling, and resource management, with upskilling available via ISHRAE and ASHRAE e-learning portals.

2.1.1 Digital Tools for HVAC Project Managers

Modern HVAC projects benefit from a wide range of software tools and platforms. Digital dashboards (e.g., Smartsheet, Primavera, BIM 360, MS Project, Zoho Projects, and Excel trackers), mobile apps for punch lists, and cloud-based document control ensure transparency, track progress, and improve communication.

Case Study: Balancing Technical & Stakeholder Needs
During the execution of a commercial office building project in Chennai, the HVAC contractor faced a dispute between the ducting team and the false ceiling contractor regarding installation sequencing. The Project Manager intervened, reviewed the drawings with both parties, and adjusted the work sequence to avoid rework. By proactively resolving the conflict, the PM maintained the project schedule and ensured quality standards were met.

Takeaway: An effective HVAC Project Manager not only understands the technical details but also facilitates smooth collaboration between trades. This case study demonstrates the qualities of a project manager to "listen" to find a solution and collaborate with all the stakeholders. A strong project manager does not just "monitor work", but actively plans, coordinates, solves problems, and leads people to achieve project goals.

2.2 Key Responsibilities of PM

The project manager is expected to coordinate with the architects, project management consultants, HVAC consultants, manufacturers, and develop contractors, work with multiple agencies such as civil, false ceiling, electrical, firefighting, etc., and different team members in the project environment.

A brief description of the key responsibilities of the project manager is enumerated below in Table 2: PM Responsibilities.

An HVAC Project Manager's responsibilities extend across all phases of the project lifecycle:

Table 2.2 PM Responsibilities

Responsibility	Description
Planning & Scheduling	Translating design intent into a practical execution plan, with milestones and dependencies clearly defined. Participate in Contract review, Design Reviews and approvals, technical submittals, Method statements, procurement planning, quality planning, acceptance criteria, applying relevant standards, communication planning and Coaching. Identify project risks, technical risks, and prepare mitigation plans to minimise the impact.
Resource Allocation	Site preparation and management, managing labour/contractors, materials, and tools to avoid idle time and cost wastage. Develop innovative methods to automate labour-oriented work and minimise. Assess the resources, capabilities, train and motivate the team.
Budget Management	Preparing, monitoring, and controlling the budget to prevent overruns. Maintaining cash flow is a KEY ATTRIBUTE for the success of a project. Understanding of relevant tax structures, supply chain costs and other commercial conditions helps the project manager to make informed decisions.
Risk Management	Anticipating risks such as delayed deliveries, site access issues, or design discrepancies, approvals, scope creep, cost overruns and preparing mitigation strategies. Risk assessment is not limited to the planning stage; it flows through every stage of the project life cycle.
Quality Assurance/ Quality Control	Ensuring work meets ISHRAE, ASHRAE, NBC, Green Buildings and other Client-specific standards as per the contract. Quality audits are performed as per the QA Plan. All inspection and testing carried out as per ITP.

Flawless Execution	Ensuring the project is carried out exactly as planned, with all technical objectives achieved. Subsystems and the entire system must perform as per design intent, delivering comfort, efficiency, and safety in real-world operation.
Safety Management	Enforcing safety protocols, statutory compliance, HSE compliance and PPE usage.
Client Communication	Maintaining clear, transparent, and timely updates to clients and consultants.
Commercial Activities	Planning, monitoring and controlling the accounts receivable and Accounts payable. Managing the proper cash flow for the projects.

Many projects fall into cost overruns and schedule overruns as the project manager does not have any clarity on the Cash flows

2.3 Skills and Competencies

The HVAC industry is more of an engineering-driven industry. Technical professionals with a technology/engineering background are the prime movers are these projects. They rise in their careers with more expertise and demonstrated excellence in technical knowledge. When some of these professionals assume the role of a project manager, their focus is more on finding a technical solution. Their lack of project management skills, resource management skills, collaboration skills and other soft skills becomes a bottleneck in their path.

A competent HVAC Project Manager requires a balanced mix of technical, project management, and leadership skills. To succeed as an HVAC Project Manager, one needs more than theory — a toolkit of practical skills. The most critical ones are outlined below.

2.3.1 Technical Skills

- Strong foundation in HVAC fundamentals, including heat load calculations, equipment sizing and selection, fan laws, pump curves, and principles of heat transfer.
- Expertise in system components such as ducting, piping, insulation, controls, and balancing.

- Proficiency in commissioning, testing, and troubleshooting HVAC systems to ensure design intent is achieved.

- Working knowledge of associated electrical systems — including panels, cabling, starters, and protective devices — to ensure seamless integration with mechanical systems.

- Ability to read and interpret shop drawings, specifications, and BOQs.

- Familiarity with ISHRAE standards, ASHRAE standards, NBC codes, relevant IS codes, and environmental compliance norms.

2.3.2 Project Management Skills

- Proficiency in scheduling tools (MS Project, Project Libre or equivalent).

- Skills in using AutoCAD, Rivet and other BIM tools.

- Project Monitoring and controlling.

- Technical submittals preparation, method statements.

- Knowledge of budgeting and cost control methods.

- Risk identification and mitigation.

2.3.3 Leadership Skills

- Leadership that motivates teams and gains trust.
- Listening skills, Empathy and communication skills.
- Emotional Intelligence.
- Negotiation and conflict resolution with contractors and vendors.
- Adaptability under changing project conditions.

2.3.4 Stakeholder Management

The HVAC project manager interacts with multiple parties:

- Clients & Consultants – Aligning on scope, approvals, and expectations.
- Contractors & Suppliers – Ensuring timely delivery, resolving supply bottlenecks.
- Internal Teams – Supervisors, engineers, and site execution teams.

2.3.5 Contract Management

A very critical skill a Project manager must master is Contract Management

- Understand contract clauses and their impact.

- Commercial knowledge with respect to taxes, shipping charges, insurance, and ownership transfer during transit.
- Interpretation of legal terms and conditions

2.3.6 Decision-Making Under Constraints

HVAC projects in India often face unforeseen challenges:

- Example 1: An AHU shipment is delayed due to a port strike. The PM evaluates alternate local suppliers, negotiates expedited manufacturing, and revises the installation sequence to keep progress moving.
- Example 2: Retrofit work can be done only during the planned shutdown for a specific time period. PM plans all the required materials and tools for the site before the start of the shutdown. Mobilises additional manpower to work and complete the tasks in the given time.

Decision-making framework: Identify → Assess → Decide → Act → Review.

2.3.7 Ethics, Safety, and Compliance

- Never compromise safety for speed, even under client pressure.
- Maintain transparency in cost and progress reporting.
- Ensure compliance with Indian labour laws, pollution control regulations, and local building codes.

Tip: Zero tolerance on Safety processes

2.4 Top 5 Mistakes HVAC PMs Make

1. **Rushing Without Approvals**
   Installing ducts or equipment before drawing/consultant approval → guaranteed rework.

2. **Weak Stakeholder Communication**
   Avoiding difficult conversations with clients, consultants, or contractors only delays conflict resolution.

3. **Compromising on Safety**
   Pressuring teams to "just finish quickly" instead of enforcing PPE, scaffolding, and safety checks invites accidents.

4. **Ignoring Cash flow**
   Many PMs track only progress, not payments. Poor cash flow planning leads to stalled work and unpaid vendors.

5. **Over-focusing on Technical Details**
   Spending hours on coil selection while ignoring scheduling, coordination, and manpower bottlenecks.

> **Golden Rule:** *"Most HVAC project failures don't come from technical errors — they come from managerial blind spots."*

2.5 Conclusion

The HVAC Project Manager is the conductor of the orchestra — ensuring every trade, supplier, and stakeholder plays in harmony. The PM's effectiveness determines whether a project is delivered smoothly and profitably, or chaotically and at a loss.

A strong PM balances technical depth with leadership, commercial acumen, and ethical responsibility. "The systems we build may run for 20 years — but the reputation we build as Project Managers lasts even longer."

Quick Cheat Sheet Role of HVAC Project Manager
Core Takeaways

- The PM is the central coordinator between design, procurement, installation, and commissioning.
- Project Management skills. Leadership, decision-making, and stakeholder management are as critical as technical knowledge.
- Balanced cash flow is a key attribute for successful HVAC project delivery.
- Safety, ethics, and compliance cannot be compromised, even under pressure.

Table 2.3 Do's & Don'ts

☑ Dos	✕ Don'ts
• Keep schedules updated and visible to all. • Build strong relationships with clients, consultants, and vendors. • Communicate delays early and proactively. • Monitor cash flows continuously.	• Don't ignore safety protocols to save time. • Don't work without a baseline schedule. • Don't avoid conflicts — resolve them with facts and collaboration. • Don't assume cash flow "will take care of itself."

Quick Tools / Must-Haves

- RACI matrix for responsibility mapping.
- Project cash flow tracker.
- Weekly progress review template.

Bottom Line

An HVAC Project Manager is not just an engineer — they are the conductor of the project orchestra. Success lies in balancing technical knowledge with leadership, stakeholder management, and financial control. In short, technical knowledge starts the journey, but it is project management excellence that delivers success.

Chapter 3: HVAC Project Life Cycle Management

"Every project has a beginning and an end — what makes the journey successful is how well we manage the steps in between."

Learning Objectives

By the end of this chapter, the reader will be able to:

1. Understand the five phases of the HVAC project management life cycle.

2. Apply life cycle thinking to align planning, execution, monitoring, and closing.

3. Recognise how HVAC-specific challenges (civil dependencies, procurement lead times, commissioning) shape each phase.

4. Use lifecycle checkpoints to prevent scope creep, rework, and handover delays.

5. Relate the PM cycle to practical site activities, approvals, and stakeholder engagement.

3.1 Introduction

In HVAC, success is measured by how smoothly the system runs after handover. Every HVAC project follows a predictable pattern, even though the details may vary.

Local Best Practices and Regulatory References

The Indian HVAC landscape is influenced by the **National Building Code (NBC)**, **ECBC (Energy Conservation Building Code), ISHRAE and the Indian Electricity Act,** which underpin site readiness, safety, and commissioning steps throughout the lifecycle.

Always refer to the latest notifications from the **Bureau of Energy Efficiency** and relevant state regulations.

Reflection: Embracing Change in Project Management

What was once a paper-driven field is now data-driven and technology-led. Remote collaboration tools, integrated modelling, real-time dashboards, and mobile-first communication are the realities for today's PMs. *Successful leaders dedicate time each year to learn and implement at least one new technology or digital best practice.*

> **Golden Rule**: *"A project without structured management is like a building without foundations — it may stand temporarily, but it won't last."*

3.2 HVAC Project Management Lifecycle

A successful HVAC project follows a logical, interconnected sequence of stages.

As defined by the Project Management Institute (PMI®), every successful project progresses through five key phases —

Initiating, Planning, Executing, Monitoring & Controlling, and Closing.

This book adopts the same structured framework, customised for HVAC projects with clearly defined gates, deliverables, and risk considerations throughout each phase."

While the terminology may vary slightly across industries, the core framework remains consistent as recommended by PMI®.

In the HVAC context, each stage has unique demands, stakeholders, and deliverables. **ASHRAE** recommends an "Integrated project delivery and building design" (IPD). (Reference: **ASHRAE Applications Handbook**.) It offers ASHRAE guidelines and standards, offering in-depth guidance on various integrated project delivery (IPD) application requirements. IPD details the integrated approach to project management.

ISHRAE also published **a "Guidebook on Integrated Approach to Project Delivery" (IAPD)** in 2016. It is a very detailed framework with templates and guidelines to execute HVAC projects and set the expectations among all the stakeholders, including Owners, facilities managers, HVAC Consultants, System integrators, equipment manufacturers, contractors, Architects, operations, maintenance teams, and other stakeholders. Roles and responsibilities are clearly defined for a successful energy-efficient project completion.

ISHRAE's Integrated Approach to Project Delivery (IAPD) book includes formats, but the scope spans broader building engineering; a focused HVAC-PM kit for Indian sites with filled examples (kick off packs, risk registers, EVM dashboards, ITPs, RFI/change logs, Cx/TAB sheets) directly usable by project engineers and supervisors is a practical differentiator.

The proposed approach in this book is a complementary methodology to these frameworks and approaches. This book is organised to cover the full lifecycle—initiation to commissioning and handover—around PM processes, decisions, and controls (scope, schedule, cost, risk, change, quality), then connects them to HVAC tasks, creating a unique "PM-as-the-operating-system" approach.

In HVAC projects, each stage has **unique demands, stakeholders, and deliverables**. Below is a structured framework tailored for HVAC projects.

3.3 HVAC Project Initiating Phase

3.3.1 Inputs:

Initial Risk Register, Organisational Assets, RFP/contract, OPR draft, constraints, standards list ECBC/ISHRAE/ASHRAE.

3.3.2 Key Activities

- Capture **Owner's Project Requirements (OPR)**: comfort cooling, process cooling, cleanroom, hospital-grade air quality.

- Coordinate with **architects & designers** on early layouts.

- Conduct **site assessments**: space availability, load-bearing capacity, existing utilities.

- Develop **feasibility & concept options**: VRF, chilled water, ducted splits, etc.

- Identify **stakeholders**: architects, consultants, contractors, vendors, and facility managers.

- Conduct a Project kick-off meeting with all the stakeholders.

- Draft **preliminary scope, budget, and schedule**.
- Identify and assign a **project manager (PM), Authorise PM**
- Review the contract for any deviations
- Prepare a Project charter

References: "NBC, ECBC, ISHRAE, ASHRAE, and the Indian Electricity Rules" and "latest notifications from BEE and state authorities."

3.3.3 Outputs

Project Charter, OPR, Kick-off meeting Minutes, Preliminary Scope Statement, Stakeholder List, Draft Contract/Abstract, Long Lead time items list, high-level schedule/cost, initial risk register.

3.3.4 Common Risks

- Misunderstood client expectations (e.g., IEQ standards).
- Underestimated space or load requirements.
- Key stakeholders were not identified early.
- Contract review was not done thoroughly, along with the detailed commercial conditions.

3.4. Project Planning Phase

3.4.1 Inputs

Kick off meeting (KOM) Minutes, Contract review, WBS, drawings, BOQ, technical submittals, method statements, scheduling, procurement planning, risk/HSE/resource planning

3.4.2 Key Activities

- **Contract Review**: Highlight key clauses and risks (LDs, warranties, BOQ ambiguities).
- Prepare **tender/shop drawings**, **BOQ**, and get consultant approvals.
- Develop **Work Breakdown Structure (WBS)** to define scope.
- Submit **technical submittals** (datasheets, compliance) for approval.
- Prepare **method statements & Inspection and Test Plans (ITPs)** for ducting, piping, insulation, and pressure tests. These method statements should be in line with the **ISHRAE, ASHRAE, and RATA** standards and should include any specific needs of the project as per the tender specifications and scope documents
- Build a **realistic schedule** (**Gantt chart**, dependencies, critical path). The schedule consists of preparing Gantt charts, identifying critical paths, and aligning with other contractors/agencies' schedules. **Be realistic in preparing the timelines**. Padding up timelines will only add more risk to the project. Get this Schedule aligned with the overall project schedule of the main project
- Prepare **procurement log** with vendor approvals, FAT, shipment, and delivery milestones. Update cash flow statements with procurement plans

- Define **resource plan** (engineers, supervisors, subcontractors).

- **Prepare Risk assessment** plan— identifying technical, commercial, and environmental risks and preparing mitigation plans.

- Prepare **HSE plan** and allocate safety resources.

- Develop **cash flow curve** and **earned value measurement (EVM) baseline**.

Outputs

Approved Project Management Plan, Drawings, baselined schedule & budget; Technical Submittals, Method Statements, Procurement Log, Resource Plan, RACI; **Risk Register**, HSE Plan, Baselined Schedule & Budget. Change Control Board (CCB) charter.

Common Risks

- BOQ mismatches with specifications.

- Long-lead imports not factored into schedule.

- Padding timelines → risk of complacency.

3.5 Project Executing Phase

Execution is where plans come to life on-site. In HVAC projects, this is not just about installing equipment — it's about precise coordination, overcoming multiple constraints and challenges without compromising the quality of the project. A proactive and agile project manager and team may find it less difficult compared to an ordinary project manager and team.

Key Activities

- Procure and store **materials & equipment**; maintain inventory traceability.
- Ensure **site readiness** (scaffolding, foundations, plant room space).
- Install **systems**: AHUs, chillers, VRFs, ducts, piping, insulation, cabling.
- Prepare for commissioning (pre-functional checks, FPT scheduling, TAB readiness) and execute as per the approved Cx plan—commissioning is a process, not a switch-on event."
- Coordinate with **civil, electrical, plumbing, and firefighting** teams.
- Ensure **QA/QC inspections** during installation.
- Conduct **Testing, Adjusting, and Balancing (TAB)** for both air and water systems.
- Enforce **safety protocols** (PPE, permits-to-work).

Outputs

Installed Systems (as per approved drawings, submittals, and method statements), updated Site Logs, and Quality Inspection Records.

Common Risks

- Delayed site readiness → idle resources.
- Poor storage → material damage.
- Misalignment with other trades → rework.

3.6 Project Monitoring & Controlling Phase

This stage runs in parallel with execution and planning, ensuring the project stays on course.

Key Activities

- Track **progress vs. baseline schedule**.
- Update **issue log** and resolve bottlenecks.
- Conduct **quality inspections** (duct leak tests, piping pressure tests).
- Monitor **cash flow & EVM metrics** (SPI, CPI).
- Manage **scope changes** via Change Control Board (CCB).
- Update **risk registers** regularly.
- Escalate unresolved issues via the **escalation matrix**.

Outputs

Updated Schedules, Variance Reports, Issue Logs, EVM Reports, Quality Inspection Checklists, Risk Updates, and Change Requests.

Common Risks

- Delayed approvals are causing cascading delays.
- Hidden costs not reflected in BOQ reconciliation.
- Incomplete quality documentation → disputes later.

3.7 Project Closing Phase

Key Activities

- **Complete Technical Closure**
- Prepare **snag list** and close items before occupancy.
- Finalise **measurements** and deviation statements.
- Handover **O&M manuals, warranties, and as-built drawings**.
- Train client's **O&M team**.
- Complete **commercial closure**: invoices, retention release, bank guarantees.
- Finalise **contract closure**: technical, commercial, legal obligations.
- Conduct **closing meeting** and document **lessons learned**.

Outputs

Commissioning Report, Snag Closure List, O&M Manuals, Final Deviation Statements, Contract Closure Report, Lessons Learned Document.

Common Risks

- Incomplete measurements → delayed payments.
- Missing warranties/test certificates → withheld bills.
- Client's O&M team untrained → poor system operation.

3.8 Lifecycle Stage-Gates (Go/No-Go checks)

- **Gate 1 – Initiating to Planning:** Charter approved; OPR defined; contract reviewed; stakeholders mapped; PM formally appointed.
- **Gate 2 – Planning to Construction phase:** Baseline schedule & budget approved; WBS frozen; drawing/submittal plan agreed; risk & HSE plans signed; procurement log with long-lead path locked; change control board (CCB) set.
- **Gate 3 – Executing to Cx:** Pre-commissioning checklists complete; test points installed; calibrated instruments available; TAB plan agreed; NCRs < agreed threshold.
- **Gate 4 – Cx to Handover:** FPT/TAB passed; documentation complete (O&M, as-builts, warranties); training done; snag list closed.
- **Gate 5 – Handover to Closing:** Technical Closure, Commercial closure (final bills, retentions, BGs), lessons learned logged, client sign-off.

3.8.1 Phase KPIs (simple, measurable)

- **Initiating:** Charter sign-off ≤ 10 working days; OPR completeness score ≥ 90%.

- **Planning:** % drawings/submittals approved on first pass; % long-lead items with FAT date locked; baseline freeze by target date.
- **Executing:** Right-first-time inspections ≥ 95%; HSE TRIR = 0; % work fronts with signed readiness checklists.
- **Monitoring &Controlling:** SPI ≥ 0.95; CPI ≥ 0.95; open issues > 14 days = 0; NCR closure SLA met ≥ 95%.
- **Closing:** Snag closure ≤ agreed SLA; O&M completeness = 100%; training attendance ≥ 95%; retention release within contract SLA.

Golden Rule: *"A beautifully designed and executed HVAC system can fail in the client's eyes if project management is weak. PM is the invisible backbone of successful delivery."*

3.9 Conclusion

HVAC projects succeed when lifecycle discipline turns moving parts into a managed system. Gates convert uncertainty into decisions, KPIs convert progress into facts, and documentation converts effort into trust. Do the right work at the right phase—and protect each handoff with clear criteria—and you will trade firefighting for foresight. The chapters ahead unpack each phase with checklists, trackers, and cases so you can run the lifecycle with confidence on your next site.

When rigorously applied, it ensures that:

- Stakeholder expectations are captured early,
- Resources and risks are planned realistically,
- Execution is monitored with facts, not assumptions,
- Closing is structured, not rushed.

Golden Rule: *"HVAC project success is not a result of luck or heroics — it is the outcome of disciplined lifecycle management."*

Summary Sheet

HVAC projects are multidisciplinary and high-risk without PM.

Structured project management prevents disputes and delays.

- **Benefits:** Timely completion, cost control, quality delivery, and client satisfaction.

Table 3. 1 Phase Activities Focus

Phase	Key Activities	HVAC-Specific Focus
Initiation	Scope, contract review, kick-off	Clarify inclusions/exclusions (wall openings, cabling, BMS).
Planning	WBS, procurement, risk, and quality plans	Include long-lead items (chillers, VRFs, AHUs).
Execution	Mobilisation, installation, safety	Coordinate with civil, fire, and ceiling works.
Monitoring & Control	Track vs baseline, manage risks, QA/QC	Dashboards, progress reviews, and issue log.
Closing & Handover	Commissioning, O&M, lessons learned	Seasonal testing, training facility team.

References

- PMBOK Guide, PMI (adapted to HVAC context).
- ISHRAE Handbook
- ISHRAE "Guidebook on Integrated Approach to Project Delivery" (IAPD)
- ASHRAE Handbook
- Author's industry experience with Indian HVAC projects.

Chapter 4:
Project Initiating

"Well begun is half done." – Aristotle

Learning Objectives

By the end of this chapter, the reader will be able to:

1. Recognise the importance of initiating HVAC projects.
2. Conduct a structured contract review covering commercial and technical aspects.
3. Identify deviations early and manage them through proper documentation.
4. Assess site readiness and ensure proper preparation before mobilising.
5. Identify long-lead and import items to prevent procurement delays.

4.1 Introduction

In the context of HVAC projects, this timeless wisdom holds true. The initiating phase precedes planning and ensures the project begins on solid ground.

HVAC projects are **multi-disciplinary** by nature, involving coordination with architects, civil contractors, plumbing teams, electrical teams, and multiple vendors.

A project that starts without a structured initiation often suffers from:

- **Unclear scope boundaries** lead to disputes over extra work.

- **Delayed mobilisation** due to lack of site readiness.
- **Procurement bottlenecks** caused by late identification of long-lead items.
- **Cost escalations** arising from **commercial or technical deviations** not addressed early in the project cycle.
- Contract misunderstandings are causing cost overruns.

In India, where project conditions are dynamic, frequent design changes, delayed approvals, and labour shortages are common, a weak initiation almost guarantees **rework, strained relationships, and financial losses**.

4.2 Top Project Initiating Mistakes in HVAC Projects (India):

- **Skipping site readiness checks** (leads to idle resources, cost overruns)
- Not mapping **long-lead and import items**
- **Overlooking contract deviations**
- Failing to involve **digital reporting tools** or cloud-based document control
- Incomplete stakeholder alignment at kick-off

Case Study: Commercial Complex HVAC Project in Bengaluru

A contractor was awarded a ₹12 crore HVAC package for a large commercial complex. Instead of structuring and initiating, the team rushed into mobilisation.

- **Missed Contract Review**: Payment terms required certification from three separate consultants. Billing delays caused severe cash flow issues within three months.

- **Poor Site Readiness Check**: Chillers arrived before shafts and equipment rooms were handed over. They sat idle for six weeks, risking damage.

- **No Long-Lead Planning**: Imported cooling towers faced four weeks of customs clearance delays.

Result: Project overran by /12 months. The client levied Liquidated Damages (LDs), wiping out most of the contractor's margins.

Lesson Learned: A structured initiation could have prevented these issues. Proper contract review, readiness checklists, and long-lead planning upfront would have safeguarded margins and timelines.

4.3 What Project Initiating Really Means

The project initiation phase is **not about paperwork alone**. It is about embedding control and clarity from the start by:

- Reviewing contracts (commercial + technical).

- Assessing site readiness using objective checklists.

- Authorising the project manager to perform the tasks

- Highlighting long-lead and imported items for procurement teams.

- Setting inspection and quality protocols before execution.

- Mapping high-level risks and mitigation measures.

A well-structured initiating phase creates alignment, minimises surprises, and ensures the project moves into planning with **confidence and preparedness**.

4.4 Kick-Off Meeting

A **kick-off meeting** formalises the start of the project, bringing together owners, consultants, contractors, and suppliers. It sets the tone and ensures everyone is aligned.

Inputs for the Kick-Off Meeting:

- **Project Scope Document** – HVAC requirements, deliverables, and boundaries (in line with contract documents, agreements, FIDIC/ISHRAE clauses, etc.).
- **High-Level Schedule** – milestones, WBS, and deadlines for HVAC activities.
- **Stakeholder List** – roles, responsibilities, and contact details.
- **Resource Plan** – manpower, equipment needs, and specifications.
- **Knowledge Repository** – lessons learned, risk logs, and reusable templates from past projects.

Kick-Off Meeting Process:

1. **Welcome & Introductions** – establish rapport and clarify roles.
2. **State Objectives** – align stakeholders on project goals.
3. **Authorise Project Manager (PM)** – formally introduce PM, assign authority, and define roles.

4. **Establish PMO (if required)** – Set up project support office for dashboards, communication, and coordination.

5. **Review Contracts & Scope** – highlight key technical, commercial, and regulatory requirements.

6. **Discuss Schedule & Milestones** – align with ISHRAE, NBC, and regulatory codes.

7. **Stakeholder Mapping & Communication Protocols** – finalise reporting formats, escalation paths, and frequency.

8. **Risk Identification** – list high-level risks (logistics, approvals, safety).

9. **Clarify Deliverables & Next Steps** – assign immediate actions.

10. **Document Minutes** – circulate to all stakeholders.

Expected Outcomes:

- **Shared Understanding** – everyone aligned on HVAC goals, scope, and constraints.

- **Project Charter Issued** – formally authorising the project and budget.

- **Defined Roles & Responsibilities** – clarity for each stakeholder.

- **Aligned Communication Protocols** – reporting and escalation in place.

- **Clarity on Milestones** – agreement on deadlines and deliverables.

- **Concise Action Plan** – immediate tasks with owners.
- **Documented Minutes** – an accurate record of discussions and risks.

Table 4.1 Do's & Don'ts

Dos	Don'ts
• Prepare thoroughly – Review scope, contract, drawings, risks, and approvals before the meeting. • Invite all stakeholders – client, consultants, contractors, subcontractors, vendors, and safety officers. Missing voices cause delays later. • Set clear objectives – Define what the project must achieve, not just when it must finish. • Document everything – Circulate detailed minutes with agreed actions and owners within 24–48 hours. • Define escalation paths – Ensure clarity on "who resolves what" at different levels. • Address site readiness upfront – Access, power, water, storage, safety protocols. • Highlight long-lead items – Discuss procurement plan for chillers, AHUs, cooling towers, VRFs. • Assign responsibilities – Publish RACI matrix for scope, approvals, and communication. • End with clarity – Make sure everyone leaves knowing their next steps and deadlines.	• Don't skip key stakeholders – Excluding subcontractors or safety officers is a recipe for rework. • Don't rush through contracts – Overlooking payment terms or LD clauses creates major cash flow problems. • Don't allow vague commitments – Every action must have a name, owner, and deadline. • Don't focus only on the schedule – Discuss risks, quality, safety, and approvals too. • Don't assume silence means agreement – Actively confirm buy-in from each participant. • Don't overload with technical details – Kick-off is about alignment, not solving every drawing clash. • Don't skip risk discussion – Identifying risks late causes surprises that could have been prevented. • Don't delay minutes – If MoM is not shared promptly, actions will be forgotten or misinterpreted.

Golden Rule: "A kick-off meeting is not about starting work — it is about starting alignment.

4.5 Contract Review in HVAC Projects

Every HVAC project must start with a **detailed contract review**. During the tender stage and subsequent sales negotiations, a lot of discussions take place, and some of these override the original clauses in the tender specifications and commercial conditions. If these are not captured systematically, they are forgotten, and the project team ends up facing surprises during execution.

Golden Rule: *"If it's not documented, it doesn't exist."*

4.5.1 Impact Analysis Document:

The contract review should result in an **Impact Analysis Document**, clearly listing:

- What is **in scope**
- What is **out of scope**
- Potential risks to cash flow, pricing, or schedule

If any major deviations exist that could impact financials or execution, they must be escalated to senior management before project initiation.

4.5.2 Commercial Aspects

- **Payment Terms:** Clearly understand advances, running bills, retention money, and liquidated damages (LD) clauses.

- **Guarantees:** Bank guarantees, performance guarantees, retention period.
- **Taxes & Duties:** GST, import duties, exemptions (if any). For imports, identify the **point of ownership change** and account for insurance/transportation costs.
- **Project Completion Date:** Confirm assumptions and penalties linked to completion.

- **Billing Procedures:** Payment terms may look fair, but if the process involves multiple certifications (client → consultant → finance), delays in billing could hurt cash flows.

- **Dispute Resolution:** An Escalation matrix for disputes should be agreed upon upfront.
- **Variations & Claims:** BOQs are never 100% accurate. Contracts that defer variation claims until the project end can create heavy losses. Negotiate progressive approvals and amendments.
- **Measurements:** Agree on industry-standard methods for ducting, piping, insulation, and cabling. Document this to avoid disputes later.

4.5.3 Technical Aspects

- **Design Intent vs Tender Specs:** Review all communications; highlight gaps between design intent, specifications, and BOQ.

- **Drawing/BOQ Gaps:** Tender drawings often differ from Good-for-Construction (GFC) drawings. Analyse possible quantity variations early.

- **Deviations:** Document deviations identified during bidding vs post-award.

- **Missing Scope Items:** Common gaps include cabling, earthing, sensors, control wiring, insulation, supports, testing, and commissioning accessories.

- **Interface Points:** Clearly define responsibilities at handover points with civil, electrical, architectural, and structural teams.

4.5.4 General Aspects

- **Site Access & Permissions:** Many clients have strict site-entry protocols. Welding, scaffolding, electrical works, or working at heights may require special permits. Non-compliance can lead to expulsion.

- **HSE Requirements:** Safety compliance is **zero tolerance**. Confirm PPE, training, and manpower certifications.

- **Site Preparation:** Identify needs for worker accommodation, sanitation, food, temporary offices, storerooms, and meeting rooms.

- **Safety Officers:** Assess the number and qualifications of certified safety officers needed.

- **Utilities:** Confirm availability of power, water, and sanitation facilities before mobilisation.

Practical Tip:

Create a **Contract Review Matrix** that captures all deviations (technical and commercial) and circulate it for client/consultant acknowledgement. This avoids disputes later and becomes part of the project's initiation record.

4.6 Identification of Tools and Resources

- Specialised tools required (duct lifters, hydraulic crimping tools, balancing instruments, welding machines).
- Calibration and certification needs before mobilisation.
- Identifying resource gaps vs subcontractors/ vendors' support.

4.7 Site Preparation & Readiness Assessment

- Site access (approvals, gate passes, crane movement areas).
- Storage areas for equipment and consumables.
- Utilities available (temporary power, water).
- Safety facilities (first aid, PPE compliance).
- Interface readiness (civil fronts, slabs, shafts, equipment pads handed over).
- Developing a **Site Readiness Checklist** before mobilisation.

4.8 Long-Lead and Imported Items

- Chillers, AHUs, cooling towers, pumps, control panels, VFDs.
- Import procedures, documentation, duties, and customs clearance.
- Coordination with international vendors for shipping schedules.
- Impact on the project baseline if delays occur.

4.9 Quality Inspection Requirements

- Early identification of inspection and test plans (ITPs)
- Factory inspections for critical equipment (chillers, AHUs).
- Third-party inspections for safety-critical items (pressure vessels, valves).
- Site-based inspections (anchor bolts, foundations).
- Record all requirements for approvals, inspections, and quality assurance tied to the HVAC installation.

4.10 High-Level Risk Identification

Conduct an initial risk assessment specific to HVAC works—consider supply chain, weather impacts, coordination with other trades, and regulatory risks.

- Contractual risks (delayed payments, scope creep).

- Procurement risks (import clearance, vendor reliability).
- Technical risks (design gaps, mismatched BOQ).
- Site risks (safety hazards, civil fronts not ready).
- Resource risks (shortage of skilled welders/TAB team).
- Prepare a Risk Register at the initiation stage that will continue to the closing stage.

4.11 Deliverables of Project Initiating Phase

At the end of this phase, the following deliverables will be prepared.

- Kick-off meeting Minutes
- Signed-off Contract **Review Report** (commercial & technical deviations).
- Site Readiness Checklist approved by client/consultant.
- Long-lead/Import Item Tracker initiated.
- Preliminary Risk Register
- Mobilisation Plan for tools, manpower, and subcontractors.

4.12 Conclusion

Project initiation is not just a ceremonial meeting — it is the **launchpad** that determines whether an HVAC project will glide smoothly or descend into chaos. Every unresolved ambiguity in scope, contract terms, or site readiness at this stage **multiplies**

downstream into disputes, delays, rework, and financial strain.

In the Indian HVAC context, too many projects rush into execution without aligning stakeholders, clarifying deliverables, or checking dependencies. The result is predictable: idle labour, delayed payments, wasted materials, and commissioning nightmares.

A disciplined initiation phase transforms this vulnerability into strength by:

- Building early trust with clients through clarity and professionalism.
- Equipping the Project Manager with foresight to anticipate risks and plan buffers.
- Preventing rework by ensuring scope, drawings, and responsibilities are frozen upfront.
- Saving months of frustration and cost by addressing problems before they reach the site.

When initiation is done well, it sets the tone for the entire project lifecycle — enabling smoother execution, stronger relationships, and higher profitability.

Golden Rule: *"Well-begun projects don't guarantee success, but poorly begun projects almost always guarantee failure."*

4.13 Summary Sheet – Project Initiating

By embedding contract review, stakeholder alignment, kick-off meetings, and readiness checks into the culture of every project, HVAC professionals can ensure they start on the front foot — making the rest of the lifecycle far more manageable.

Purpose: Convert an idea into a formal, structured project.
Key Deliverables: Project Charter, Feasibility Study, Stakeholder Map, Preliminary Risk Register.

Golden Rule: "Weak initiation = strong firefighting."

Table 4.2 Project Initiating Activities Summary

Step	Key Actions	Deliverables
1. Contract Review	Review technical & commercial terms; list scope inclusions/exclusions; highlight billing risks & long-lead items.	Contract Review Matrix
2. Stakeholder Alignment	Map all stakeholders, define roles/responsibilities, set escalation matrix & communication protocols.	Stakeholder Map & RACI
3. Kick-off Meeting	Conduct a formal meeting; align objectives, schedules, milestones, and reporting formats; circulate MoM.	Kick-off MoM
4. Site Readiness Check	Verify site access, permits, utilities, foundations, and civil handover before dispatching equipment.	Site Readiness Checklist
5. Risk Identification	List foreseeable risks (procurement delays, labour, regulatory); capture in preliminary register.	Preliminary Risk Register
6. Project Charter	Document objectives, budget, high-level schedule, and authority of PM; obtain sponsor sign-off.	Project Charter

References

- PMBOK Guide (PMI®) (adapted to HVAC context).
- ISHRAE Handbook
- ISHRAE "Guidebook on Integrated Approach to Project Delivery" (IAPD)
- ASHRAE Handbook
- Author's industry experience with Indian HVAC projects.

Chapter 5
Project Planning

"Plans are worthless without action, but planning is everything." – Dwight D. Eisenhower."

Learning Objectives

By the end of this chapter, the reader should be able to:

1. **Explain the importance of Project Planning** as the foundation for successful HVAC project delivery.

2. **Develop a Work Breakdown Structure (WBS)** tailored for HVAC activities such as ducting, piping, and commissioning.

3. **Prepare and manage technical submittals, shop drawings, and method statements** for consultant/client approvals.

4. **Develop realistic schedules** that account for dependencies, long-lead equipment, approvals, and site readiness.

5. **Prepare a project budget** aligned with the schedule, covering equipment, labour, overheads, contingencies, and cash flows.

6. **Plan procurement activities**, especially for long-lead imported equipment (e.g., chillers, VRF systems).

7. **Integrate quality, risk, and other considerations** into the planning stage for seamless execution.

8. **Recognise the importance of early commissioning planning**, ensuring performance validation is not left until the end.

9. **Apply practical tools and templates** (WBS checklists, submittal trackers, procurement logs) for real-world projects.

5.1 Introduction to HVAC Project Planning

5.1.1 Importance of Project Planning in HVAC Projects

> **Golden Rule:** *"Project Plan is not a static document — it is a living activity that evolves with the project and keeps it on course."*

Project planning is often described **as the backbone of successful project management**. In HVAC projects, this holds even greater weight because of the high complexity: multiple interdependent systems, reliance on civil and electrical readiness, and the challenges of long-lead imported equipment.

A robust Project Planning **transforms the broad intent defined during initiation into a structured roadmap** that guides every stakeholder — from design engineers to site supervisors — toward project completion.

Unlike routine projects, HVAC Project planning must synchronise **design details, procurement, subcontractors, manpower, testing, commissioning, and client approvals** into a single, Integrated Project Plan. (IPP)

5.1.2 Why Planning Is Critical in HVAC Projects

- **Clarity of Scope** – Translating tender documents into a Work Breakdown Structure (WBS) avoids ambiguity about inclusions and exclusions.
- **Realistic Scheduling** – A strong plan aligns civil handovers, equipment lead times, approvals, and testing into a logical sequence.
- **Resource Optimisation** – Manpower and subcontractors are allocated based on project peaks and valleys, preventing idle crews or sudden shortages.
- **Financial Discipline** – A cost baseline and cash flow forecast allow both contractor and client to anticipate funding needs and avoid disputes.
- **Risk Mitigation** – Potential challenges like import delays, design changes, or labour disruptions can be foreseen, with mitigation strategies in place before they escalate.

Practical HVAC Example

A seemingly small oversight — failing to align **ducting and insulation work with false ceiling schedules** — can trigger cascading effects: duct rework, idle resources, delayed ceiling works, increased cost, and strained client relationships.

In the **Indian context**, where projects often face uncertainties such as fluctuating delivery times, uneven labour productivity, and consultant-driven approval delays, robust planning is not a ***"nice-to-have"*** — it is the **difference between a project that**

delivers smoothly and one that bleeds time and money.

A study shows that more than 75% projects plan only for a Gantt Chart with materials planning and a few milestones.

Case Study: IT Park HVAC Project – Success Through Meticulous Planning

A large IT park project in Bengaluru involved the installation of central air-conditioning systems for five towers, each with multiple floors and diverse tenant requirements. The contractor faced several challenges upfront: long-lead imports (chillers from China), strict deadlines to align with the phased handover of towers, and multiple subcontractors working simultaneously at the site.

Instead of rushing into execution, the project team devoted six weeks to a detailed integrated planning of all contractors' activities.

- **Comprehensive WBS**: The scope was broken into systems (chillers, AHUs, ducting, piping, controls) and mapped against civil progress.
- **Realistic Schedule:** Dependencies with slab casting and shaft openings were built into a detailed CPM schedule.
- **Procurement Strategy**: Long-lead equipment orders were placed early, with vendor progress tracked weekly. Alternate suppliers were pre-identified in case of delays.

- **Resource Planning**: A resource histogram ensured peak manpower demand during ducting and piping phases was met without overstaffing earlier.

- **Risk Mitigation**: Import clearance was flagged as high risk. The team appointed a dedicated logistics consultant to handle customs proactively.

Result:

- All towers were completed on schedule, with HVAC systems handed over in phases exactly as per client commitments.

- No penalties were levied, and the client expressed satisfaction with the predictability of progress.

- The project team's reputation improved significantly, leading to repeat orders for subsequent IT projects.

This case demonstrates how disciplined planning—covering schedule, procurement, resources, and risks—creates certainty in delivery, even for complex projects in the Indian context.

5.1.3 Project Planning Basics:

PMI[1] recommends different knowledge areas to be considered for planning. If one follows this framework, the probability of the project's success increases.

Those are explained in detail in PMBOK.

Detailed and careful planning of these is very important in HVAC projects.

- Scope management
- Schedule management
- Cost Management
- Quality Management
- Integration Management
- Procurement Management
- Resource Management
- Risk Management
- Communications Management and
- Stakeholder Management

A complete "Integrated Project Plan"(IPP) is prepared to include all the above areas in detail. IPP is like a roadmap for the entire project with all the necessary tools and techniques required to steer the project.

"A Schedule is not a complete project plan. The schedule is only a subset of a complete project plan.

HVAC Integrated Project Planning Flow diagram is depicted in Fig 1, Project Planning Flow below:

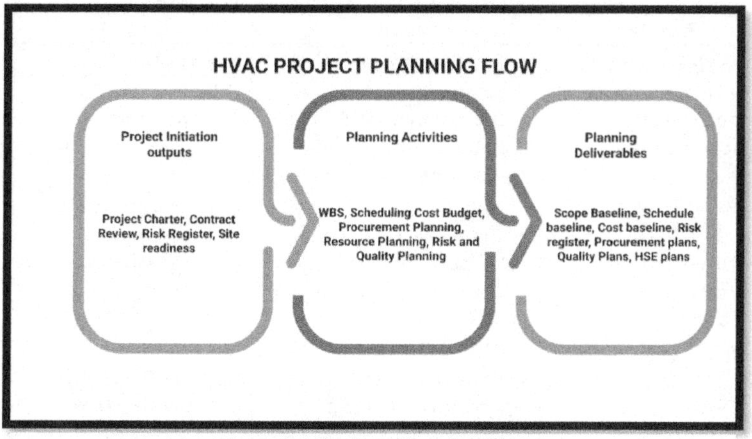

Figure 5.1 Project Planning Flow

5.1.4 Project Planning Framework (Inputs → Process → Outputs)

A structured planning framework helps project managers move from **broad intent** to **detailed, actionable plans**. In HVAC projects, where dependencies and risks are high, this clarity is essential.

Table 5.1 **Project Planning Framework**

Inputs	Process	Output
• **Project Charter & Initiation Documents** – Stakeholder map, preliminary risk register. • **Contract Documents** – Scope, technical specification, BOQ, commercial conditions. • **Design Basis Report (DBR) & Tender Drawings** – To be developed into shop drawings. • **Standards & Codes** – ISHRAE, ASHRAE, NBC, IS codes, SMACNA. • **Site Readiness Data** – Civil progress reports, power availability, site access conditions. • **Client/Consultant Requirements** – Submittal timelines, approval processes, special conditions	• **Define Work Breakdown Structure (WBS)** – Break scope into manageable activities. • **Prepare Detailed Schedules** – Using Gantt, CPM, PERT, or CCPM techniques. • **Procurement Planning** – Map long-lead items (chillers, VRFs, pumps) with milestones. • **Budget & Cash Flow Forecasting** – Establish cost baseline and cash flow curves. • **Quality Planning** – Inspection & Test Plans (ITPs), acceptance criteria, QA/QC roles. • **Risk Planning** – Identify risks and mitigation measures. • **Commissioning Planning** – Define early testing and handover activities. • **Communication & Stakeholder Plan** – Define reporting, meetings, and escalation matrix.	• **Baseline Schedule** – Approved, frozen schedule (v1.0). • **Procurement Tracker & Milestones** – Linked to delivery/installation. • **Cost Baseline & Cash Flow Plan** – For financial discipline. • **Quality Assurance Plan (QAP) & ITPs** – With checklists and templates. • **Risk Register (detailed)** – With mitigation, owners, and response strategies. • **Commissioning Roadmap** – Milestones for testing, balancing, and seasonal performance. • **Communication Plan** – Meeting formats, reporting dashboards, escalation paths.

Golden Rule: *"If the inputs are unclear, the plan will fail — planning is only as strong as the clarity of information at the start."*

5.2 Scope Planning and Work Breakdown Structure (WBS)

The **Work Breakdown Structure** (WBS) serves as the foundational tool for scope management in HVAC projects by systematically converting the broad contract scope and Bill of Quantities (BOQ) into detailed, actionable deliverables. By breaking down the overall project scope into smaller, manageable components or work packages, the WBS provides clarity on exactly what needs to be delivered and facilitates clear assignment of responsibilities. This structured decomposition not only helps in accurately defining the project's boundaries but also forms the basis for subsequent planning activities such as scheduling, cost estimation, resource allocation, and risk management. In essence, the WBS transforms the often complex and high-level contractual requirements into a practical blueprint that guides project execution with precision, ensuring that all necessary HVAC tasks—from duct fabrication to equipment commissioning—are clearly accounted for and controlled throughout the project lifecycle. The plan also includes how Scope changes are managed during the project execution stage.

5.2.1 Importance of WBS in HVAC Projects

It is necessary to understand the importance of preparing a detailed work breakdown structure.

A clear WBS is like a project's DNA — if defined correctly, everything else falls into place."

A complete and structured WBS has the following advantages.

  - Provides a clear map of all deliverables.

- Avoids ambiguity between client, consultant, and contractor.

  - Facilitates scheduling, budgeting, and resource planning.
  - Enhances accountability by assigning ownership at each level.
  - Serves as the baseline for monitoring and control.

5.2.2 Levels of WBS

A typical WBS for an HVAC project has 3–5 levels, depending on project complexity.

Level 1 – Project

  - Complete HVAC project (e.g., “HVAC Works for XYZ IT Park”)

Level 2 – Major Systems

  - Chilled Water System
  - Air Distribution System
  - Ventilation & Exhaust System
  - Electrical Panel and Cabling
  - Control & Automation (BMS/EMS)
  - Ancillary Works (insulation, supports, testing, commissioning).

Level 3 – Subsystems / Work Packages (example for Chilled Water System)

- Chillers
- Pumps
- Cooling Towers
- Piping Network
- Valves & Accessories

Level 4 – Activities

- Material submittals and approvals
- Procurement & delivery
- Installation (erection, alignment, fixing)
- Testing & inspection
- Commissioning

Level 5 – Detailed Tasks (if needed) For high-value/critical items (e.g., Chiller foundation, alignment, connection of electrical panels, vibration test).

Annexure:
Excel template for a sample WBS
Sample WBS for a Commercial Building Project

1.0 HVAC Project – XYZ IT Park

- **1.1 Chilled Water System**
    - 1.1.1 Chillers
    - 1.1.2 Pumps (Primary, Secondary, Condenser water pumps)
    - 1.1.3 Cooling Towers
    - 1.1.4 Piping
    - 1.1.5 Pipe Insulation
    - 1.1.5 Valves & Fittings
- **1.2 Air Distribution System**
    - 1.2.1 Air Handling Units (AHUs)
    - 1.2.2 Treated Fresh Air units
    - 1.2.3 Fan Coil Units (FCUs), CSUs
    - 1.2.4 Ducting (Rectangular/ Spiral)
    - 1.2.5 Insulation
    - 1.2.6 Pre-Insulated ducts
    - 1.2.4 Diffusers & Grilles
- **1.3 Ventilation System**
    - 1.3.1 Toilet Exhaust
    - 1.3.2 Kitchen Exhaust
    - 1.3.3 Basement Ventilation

- **1.4 Controls & Automation**
  - 1.4.1 Electrical Panel
  - 1.4.2 Cable Laying
  - 1.4.3 Earthing
- **1.5 Controls & Automation**
  - 1.5.1 BMS Panel
  - 1.5.2 Field Devices
  - 1.5.3 Integration
- **1.6 Ancillary Works**
  - 1.5.1 Insulation
  - 1.5.2 Supports & Hangers
  - 1.5.3 Testing & Balancing
  - 1.5.4 Commissioning & Handover

5.2.3 Deliverables of WBS

- Approved WBS (signed off by stakeholders)
- WBS dictionary (explanation of each code/work package)
- Basis for scheduling and budgeting

5.2.4 Commissioning in WBS

One of the most common mistakes in HVAC project planning is treating commissioning as a closing activity to be rushed at the end. Commissioning is a **progressive process that starts from the planning stage and continues throughout the project lifecycle.**

Why include commissioning in the WBS early?

- Ensures that system readiness is planned in line with civil and electrical handovers.
- Identifies long-lead testing instruments, tools, and third-party inspection needs.
- Allows progressive testing (hydro tests, duct leak tests, flushing) instead of leaving everything for the end.
- Guarantees compliance with design intent and contractual performance requirements.
- Prevents delays in final handover caused by missed pre-commissioning steps.

Typical Commissioning-Related Activities in WBS:

- Preparation of **commissioning plan and checklists** during the planning stage.
- Identification of test & balancing equipment and resources.
- Progressive inspections: hydro-testing of piping, duct leak testing, and electrical checks.
- Pre-commissioning activities: flushing, cleaning, alignment, and insulation checks.

- System commissioning:
    - Chillers
    - Pumps
    - Cooling towers
    - AHUs & FCUs
    - Ventilation systems
    - Controls & automation (BMS)
- Performance testing against design parameters.
- Documentation: as-built drawings, O&M manuals, training of client's staff.
- Final acceptance and handover.

"Commissioning is not the last step of a project — it begins the day planning starts."

5.2.5 Common Pitfalls in Preparing a Work Breakdown Structure (WBS)

A few common pitfalls and responses to avoid them are listed in Table 1 below.

Table 5.2 WBS Pitfalls

Description	Pitfall	Impact	Response
Missing scope Items	Important activities (e.g., insulation, testing, commissioning) are left out.	Leads to cost overruns, scope disputes, and rework.	Cross-check WBS against BOQ, tender drawings, and contract scope; use a checklist for typical HVAC activities.
Excessive Detailing	Breaking down to an unmanageable micro-level (e.g., listing each valve separately).	Creates unnecessary complexity and bloated schedules.	Keep WBS to 3–5 levels; go deeper only for high-risk or high-cost items. The cost of monitoring should not be more than the cost of the task itself, unless it is critical.
Mixing Deliverables and Activities	Including verbs/actions like "install piping" in the WBS instead of deliverables	Confuses stakeholders and blurs accountability.	WBS should describe "what" will be delivered, not "how" it will be done (activities belong to the schedule).
Ignoring Interfaces with Other Disciplines	WBS was developed in isolation from civil and electrical works.	Missed dependencies → delays when fronts are not available.	Include interface points (e.g., slab openings, cable trays) explicitly in WBS review.
Lack of Standardisation	Different projects use different WBS structures.	Hard to compare projects or build organisational learning.	: Develop a standard WBS template for HVAC projects within the organisation.
No Ownership Defined	Work packages are created but not linked to responsible persons.	Leads to accountability gaps.	Each Level-3 or Level-4 work package should have a designated owner.
Overlooking Soft Deliverables	Only physical items (ducts, pipes, chillers) are included; documents like O&M manuals, as-built drawings, and approvals are missed.	Handover delays and disputes at project closure	Always include documentation and compliance deliverables as part of the WBS.

5.2.6 WBS Quality Checklist

Before finalising your Work Breakdown Structure, ensure that:

- **Complete Scope Covered**

All contractual items (equipment, installation, testing, documentation) are included.

Cross-checked against BOQ, drawings, and tender requirements.

- **Consistent Level of Detail**

No package is over-detailed or under-detailed compared to others.

Typically, 3–5 levels deep.

- **Deliverables-Oriented**

WBS focuses on *what* needs to be delivered, not *how* it will be done.

Include all Quality Assurance Audits, Monthly Quality review meetings, Quality Control Tasks and deliverables.

- **Interfaces Identified**

Civil and electrical dependencies (openings, supports, power supply) are explicitly included.

- **Communication Deliverables Covered**

Weekly review meetings, Monthly Stakeholder meetings are better updated with a recurring task to fall on a specific day or date of the week/month.

- **Ownership Defined**

Each work package has a responsible person or subcontractor assigned.

- **Dynamic and Updatable**

The WBS is treated as a living document, updated for scope changes and variations.

5.2.7 WBS Dictionary:

A **WBS dictionary** is a complementary document to the Work Breakdown Structure that provides detailed descriptions for each work package or element within the WBS. It serves as a vital reference to clarify the scope, deliverables, milestones, and associated responsibilities for every component, eliminating ambiguity and ensuring all stakeholders have a shared understanding of what is included and excluded.

In HVAC projects, the WBS dictionary might specify technical details such as duct specifications, equipment models, testing standards, or installation methods tied to each work package. By elaborating on what each element entails, the WBS dictionary enhances project clarity, facilitates communication, improves coordination, and sets a firm foundation for scheduling, budgeting, and quality control processes.

	WBS Review Checklist		
No.	Review Item	Yes / No /	Remarks
1	WBS is fully decomposed into monitorable activities		
2	Owners are assigned for each activity		
3	Resources (manpower, tools, equipment) are assigned		
4	Activity durations are realistic and agreed		
5	Logical dependencies (FS, SS, FF, SF) are defined		
6	Procurement milestones (long-lead items, imports, FATs) are included		
7	Commissioning milestones (pre-commissioning, TAB, FPT, training, seasonal tests) are included		
8	Approvals (consultant/client, technical submittals) are explicit tasks in the schedule		
9	Risks identified during planning are considered (buffers, alternate paths)		
10	Resource conflicts checked and leveled		
11	Seasonal/environmental constraints modeled (e.g., monsoon, holidays)		
12	Schedule reviewed with all stakeholders (PM, procurement, QA/QC, consultants, client)		
13	Baseline is signed off as Version 1.0		
14	Change Control process defined for any future schedule revisions		
15	Baseline schedule uploaded/shared in common platform (PMO, site, client)		

Figure 5.2 WBS Review Checklist

Sample WBS Dictionary (Duct Fabrication)

Table 14.3 Sample WBS Dictionary

WBS Code	**1.3.1**
Work Package Name	Duct Fabrication
Description	Fabrication of all supply, return, and exhaust ducts as per approved shop drawings, including cutting, rolling, welding, insulation, and quality checks.
Deliverables	Fully fabricated and insulated ducts ready for site installation.
Responsible Party	HVAC Fabrication Subcontractor
Acceptance Criteria	Duct dimensions and materials conform to specifications; no visible leaks; insulation R-value as per design; certification from QC inspector.
Estimated Duration	20 working days
Dependencies	Receipt of approved shop drawings; procurement of raw materials.
Key Risks	Delayed material delivery; dimensional errors leading to rework.
Controls & Monitoring	Daily fabrication progress reports; weekly quality checks; deviation logs.

5.2.8 Converting WBS to a Schedule

Table 5.4 WBS to Schedule conversion summary

Step	Description	HVAC Example
WBS Definition	Hierarchical breakdown of project scope and deliverables.	Ductwork fabrication, piping installation.
Task Identification	Define activities under each WBS work package.	Material procurement, assembly, and Leak Testing.
Activity Sequencing	Establish logical order and dependencies.	Fabrication before installation
Duration & Resource Assignment	Allocate time and manpower/resources.	10 days for duct assembly with 5 workers.
Milestone Integration	Set major project checkpoints derived from deliverables.	Drawings approval, installation completion.
Schedule Development & Control	Create and update the project timeline.	Gantt chart reflecting the duct schedule

5.3 Drawings Submission and Approval Process in HVAC Projects

Drawings form the backbone of HVAC execution. Tender documents usually include suggested duct routings, equipment locations, and pipe layouts as part of the BOQ. However, before execution, these must be verified, updated, and approved through a systematic process:

5.3.1 Step-by-Step Process

1. Site Verification

- Visit the site with tender and sectional drawings in hand.
- Request the latest revisions of (Tender drawings) GFC civil/structural drawings to ensure no revisions are missed.
- Measure the room dimensions and verify beam/column layouts — these may have changed since tender.
- Verify the routings for ducts, piping, if any conflicts with beams or other obstructions.
- Record all variations and discrepancies.

2. Preparation of Shop Drawings

- Shop drawings are detailed execution level drawings to be prepared by the HVAC contractor.
- Incorporate site measurements, design intent, and heat load requirements.
- Create accurate shop drawings for submission and approval.
- Get the sizes from the original equipment manufacturers and update the plant room/AHU rooms.
- Add detailed sectional views, especially where ducts/pipes cross beams, ceilings, or other services.

- Ensure enough space around the equipment for operation and maintenance
- Include installation notes and special instructions to guide execution teams.
- ❌ Never accept incomplete or ambiguous drawings.
- Use 3D software for complex routings — early clash detection saves costly rework.

3. Reference Standards

- Cross-check shop drawings against technical specifications, BOQ, and applicable codes (ASHRAE/ISHRAE/NBC/UL, etc).
- Ensure accuracy, completeness, and standard references.

4. Take-off & Deviation Statement

- Prepare take-off quantities based on shop drawings.
- Compare with the tender BOQ and prepare a deviation statement.
- Get variations formally approved by the competent authority before procurement.

5. Approval Process

- Before submitting for approval, review thoroughly for any missing information
- Submit drawings for consultant/client approval.

- Factor in adequate time for review and revisions — delays are common.
- Track status using a *Drawing Tracker* **(sample provided in annexure).**

6. Finalisation and Communication

- Ensure all approved drawings are signed by the appropriate authority.
- Record all comments provided with the approval.
- Issuing only the latest approved version to vendors and subcontractors using outdated drawings is one of the most common causes of rework.
- Maintain a signed copy of every approved drawing at the site office/PMO for quick reference.

5.3.2 Dos and Don’ts of Drawing Preparation and Approvals

Table 5.5 Do’s & Don’ts

☑ Dos	✕ Don’ts
• Verify site dimensions, beams, and columns before drafting. • Always cross-check against BOQ, tender specs, and ISHRAE/ASHRAE standards. • Get the Measurements from OEMs for all equipment • Provide enough space for operation, maintenance and safety • Prepare deviation statements and get them approved before procurement. • Track approval status with a Drawing Tracker and build allowance for revisions in the schedule. • Circulate only the latest signed/stamped version to all stakeholders. • Keep one officially approved set available at the site/PMO office.	• Never proceed with incomplete or unapproved drawings. • Don’t assume tender drawings are final — site changes are common. • Don’t release Shop drawings verbally or over WhatsApp without formal sign-off. • Don’t ignore minor mismatches (beam depth, ceiling height) — they cause major rework later. • Don’t allow multiple drawing versions to float around; it’s the #1 cause of rework. • Don’t treat approval delays as “normal” — escalate early with evidence.

Best Practices:

- Submit all the drawings related to a zone at once.
- Do not submit piecemeal drawings.
- Ensure these drawings are thoroughly reviewed against requirements and contract conditions before submission.
- Nearly 80% of drawings are returned by consultants because they fail to meet basic requirements or formatting.
- A rigorous internal review before submission saves time for everyone.
- Follow the drawings submission and approval tracker for quick resolution and approvals

Typical Drawings Checklist, Drawings Tracker are given in the annexure.

> **Golden Rule – Drawings Approval -** *"The right drawing at the right time saves the project from endless rework. Always verify, approve, and communicate only the latest signed drawings."*

5.4 Technical Submittals in HVAC Projects

Technical submittals are a critical part of project planning in HVAC projects, yet they are often underestimated. They are formal documents prepared by the contractor to demonstrate that materials and equipment proposed for the project comply with the approved drawings, technical specifications, and Bill of Quantities (BOQ). They also meet the requirements of applicable Standards.

Submittals are not **"extra paperwork**." They are the bridge between design intent and execution reality. Without proper submittals, procurement becomes guesswork, approvals get delayed, and costly rework is almost guaranteed.

Golden Rule for Preparing Technical Submittals

Always follow this hierarchy of references:

1. **Drawings** – first point of validation
2. **Bill of Quantities (BOQ)** – second level of check
3. **Technical Specifications** – final source for detailed clarification

When in doubt, move in this order: Drawings → BOQ → Specifications.

5.4.1 Content of a Technical Submittal

A typical submittal package should include:

- Product data sheets (manufacturer catalogues, model numbers, performance ratings)
- Material specifications (gauge, thickness, coating, fire rating, acoustic rating, etc.)
- Compliance statement (clause-by-clause confirmation against specifications)

- Drawings and layouts showing integration with other services
- Certifications (energy efficiency labels, IS/ISHRAE/ASHRAE/UL compliance, green building)
- Factory test certificates (FAT reports, performance certificates)
- Installation methodology (where relevant, e.g., for chillers or VRF systems)
- Any special requirements (custom finish, warranty terms, O&M instructions)

Every submittal should reference the Drawing number, BOQ serial number, and the exact location where the material is to be installed.

The Sample Technical Submittal is attached in the annexure

5.4.2 Why Technical Submittals Matter

1. **Ensures Compliance** – Confirms that materials/equipment meet contract requirements before procurement., Maintains Traceability
2. **Avoids Rework** – Approvals upfront prevent costly changes at the site.
3. **Facilitates Coordination** – Aligns all stakeholders (consultant, client, contractor, vendor) on what will be delivered.

4. **Improves Quality** – Ensures only tested and certified materials are installed.

5. **Builds Transparency** – Documentation enhances trust with consultants and clients.

6. **Supports O&M** – Approved submittals become part of the records for maintenance teams.

5.4. 3 Do's and Don'ts of Technical Submittals

Here are some Dos and Don'ts to be remembered before making and submitting the technical submittals.

Table 5.6 Do's & Don'ts

✅ Dos	❌ Don'ts
• Ensure **submittals are complete, accurate, and reviewed internally** with a checklist before submission. • Cross-check against the **latest revision of drawings** and specs. • Highlight any **variations/deviations** with justifications. • **Submit well in advance** to allow consultant review without delaying procurement. • **Use standard formats** (cover sheet, compliance matrix, manufacturer data sheets). • Maintain a **submittal log** to track approvals, comments, and revisions.	• Don't submit **generic catalogues** without marking specific models. • Don't propose **inferior materials** without prior approval. • Don't cause **multiple resubmissions** due to careless copy-paste. • Don't ignore consultant comments — **every remark must be addressed**. • Don't delay submittals until after PO placement — **rejection here = financial loss.**

Golden Rule - *"Technical submittals are not paperwork — they are the first quality checkpoint of the project."*

5.5 Method Statements in HVAC Projects

Method statements are **vital documents** that define the step-by-step methodology for installing equipment and systems in HVAC projects. They act as a practical guide, ensuring that installations are safe, accurate, and compliant with specifications and standards.

Without method statements, young engineers and supervisors often rely solely on subcontractors, which risks misalignments, poor workmanship, or even system failures. A well-prepared method statement brings **consistency, accountability, and quality** to the project.

"A method statement is not paperwork for filing — it is a working tool. If it is not followed at the site, it has no value."

5.5.1 Key Requirements of a Method Statement

1. **Prepared by Competent Team**
   - Must be written by experienced engineers familiar with specifications and site practices.

2. **Reference Standards**
   - Must reference BOQ, technical specifications, OEM Instructions and relevant standards (e.g., SMACNA for ducting, IS/ASHRAE for piping and insulation).

3. **Comprehensive Scope**
   - Should cover:
     - Quality Assurance Plan (QAP)

- Inspection & Test Plan (ITP)
- Safety requirements (permits, PPE, scaffolding, confined space, working at heights)
- Pre-requisites before installation (civil readiness, foundation checks).

4. **Measurement Protocols**

- Define how measurements (ducting, piping, insulation, cabling) will be recorded and verified.

5. **Roles & Responsibilities**

- Specify roles of supervisors, executors, inspectors, and approvers.

6. **Documentation & Approval**

- Submit to the consultant/client for approval before starting project execution.
- Approved versions must be signed and kept at the site.

7. **Training & Implementation**

- Train site supervisors and foremen.
- Ensure copies of method statements are available on-site for reference during project execution.

5.5.2 Systems Requiring Method Statements

- **Air Distribution Systems:** Ducting, insulation
- **Hydronic Systems:** Piping (MS, copper), pumps, chillers, cooling towers
- **Special HVAC Systems:** VRF systems (indoor/outdoor), split units
- **Air Handling & Ventilation:** AHUs, fans (jet fans, centrifugal, exhaust), ventilation shafts
- **Electrical & Controls:** Panels, cabling, trays, BMS connections
- **Foundations & Supports:** Chiller foundations, equipment bases, structural supports

Sample Method statements are given in Annexure

5.5.3 Do's and Don'ts of Method Statements

Table 5.7 Do's & Don'ts

✅ DOs	❌ Don'ts
• Do prepare method statements well before starting work — not after site execution begins. • Do align every method statement with **approved drawings, BOQ, and specifications**. • Do include **safety protocols, QA/QC steps, and inspection hold points**. • Do get method statements **reviewed internally** before submitting to the consultant/client. • Do ensure the **latest approved version** is always available at the site office. • Do train supervisors and subcontractors to actually **follow the document**, not just file it.	• Don't copy-paste generic templates without customising for project-specific requirements. • Don't ignore pre-requisites (like civil readiness, foundation checks) — they must be stated clearly. • Don't proceed with installation if the method statement is **pending approval**. • Don't allow multiple versions to circulate at the site — this is a major cause of rework. • Don't skip updates — revise method statements if there are **design changes or site variations**.

> **Golden Rule – Method Statements** - *"Never begin installation without an approved method statement — it is the roadmap that transforms drawings into reality, safely and correctly."*

5.6 Project Scheduling

"A schedule is not just about dates — it is the map that tells you when your project will succeed or fail."

5.6.1 Introduction:

Why Scheduling is Crucial in HVAC Projects

"A project schedule is not just about dates on a chart — it is the roadmap that tells you when and how success will be achieved."

In HVAC projects, scheduling is the lifeline that keeps all activities aligned. From the designs of equipment, equipment procurement, to site installation and commissioning, every phase has multiple interdependencies. Unlike other projects, HVAC execution depends heavily on civil works (plant room readiness, slab openings), Site readiness to work, electrical availability (panels, cabling), and imported long-lead items (chillers, VRF systems, cooling towers). Any slip in one activity can cause a ripple effect across the project.

For example:

- If duct fabrication drawings are not approved on time, site teams may go idle, or worse, install ducts from an older version of drawings, leading to costly rework. A lot of other works, like insulation, false ceiling, plumbing, firefighting and other works, get delayed.

- In a large IT campus project, improper sequencing of piping and cabling caused clashes during ceiling installation, forcing rework and delaying commissioning by over a month.

These examples underline one truth: in HVAC projects, scheduling is not a theoretical exercise — it is the difference between smooth execution and costly delays.

It is the responsibility of the **Project Manager** to ensure all such dependencies are identified, documented, and highlighted in the schedule, along with clear roles and responsibilities of the project team.

Tip:

- The project schedule must list all activities clearly, along with their owners.
- If customer approvals, consultant reviews, or technical sign-offs are required, they should be specifically mentioned as separate activities.
- The purpose of listing activities is not to assign blame, but to ensure every stakeholder is aware of their role — and understands the impact of any delay on the entire project.

Golden Rule: *"Scheduling Dependencies don't cause delays — ignoring them does."*

Table 5.8 Scheduling Input – Process - Output

Inputs	Process	Output
o **Approved Project Charter & Scope Statement** – defines objectives and boundaries. o **Work Breakdown Structure (WBS)** – already detailed earlier; forms the foundation for breaking deliverables into activities. o **Activity Lists & Dependencies** – identifying sequencing (FS, SS, FF, SF). o **Resource Availability** – manpower, equipment, subcontractor, and their constraints. o **Procurement Plan** – long-lead items (chillers, VRFs, AHUs, pumps, cooling towers) with vendor and logistics **timelines**.	o **Define Activity Sequencing** – map dependencies between tasks. o **Estimate Durations** – realistic durations, avoiding excessive padding. o **Assign Resources & Owners** – ensure each activity has an accountable owner. o **Apply Scheduling Techniques** – Gantt (visibility), CPM (control), PERT (uncertainty), CCPM (resource reality). o **Integrate Procurement & Approvals** – include vendor lead times, FATs, customs clearance, consultant approvals.	o **Baseline Schedule** – frozen, approved plan for execution and monitoring. o **Milestone Chart** – clear visibility of critical checkpoints (procurement, installation, commissioning). o **Critical Path & Float Analysis** – identification of zero-float tasks driving completion. o **Resource-Aligned Schedule** – manpower and equipment matched to activities. o **Monitoring & Reporting Tools** – trackers, dashboards, and variance reports. **Claims & Dispute Support** – evidence for delays, variations, and extensions of time (EOT).

Inputs	Process	Output
o **Risk Register** – seasonal issues, import clearances, client approvals, and other uncertainties. **Project Assumptions & Constraints** – e.g., power availability, site readiness, access restrictions	o **Review Risks & Buffers** – add contingency or feeding buffers where required. o **Develop Draft Schedule** – circulate to team, resolve conflicts, and finalise. **Baseline the Schedule** – freeze as Version 1.0 with formal sign-off.	

5.6.2. Inputs → Process → Outputs of Project Scheduling

A project schedule is only as strong as the clarity of its inputs and the discipline of its monitoring. In HVAC projects, delays often occur not because of poor execution, but because dependencies, resources, or risks were not properly captured at the planning stage. The IPO framework ensures scheduling is systematic and transparent.

> **Golden Rule – IPO:** *"A schedule is only as good as the inputs; garbage in means garbage out. Start with clarity, end with control."*

5.6.3 Advantages of Detailed Project Schedule

A well-prepared and actively monitored schedule helps project teams as below:

1. Coordinate dependencies across other trades /agencies

- Align HVAC activities with civil, electrical, plumbing, Firefighting, interiors and finishing works.
- Avoid clashes like ductwork vs. false ceiling or piping vs. electrical trays.
- Use common services such as scaffolding and cranes effectively.

2. Plan for Long-Lead Items

- Ensure timely procurement and delivery of chillers, AHUs, pumps, VRFs, and cooling towers.
- Integrate import lead times, factory acceptance tests (FAT), and approvals into the baseline schedule.

3. Control Cost and Resources

- Proper scheduling minimises idle manpower and machinery costs.
- Avoids rush procurement and unnecessary overtime at the site.

4. Enable Timely Approvals

- Schedule drawing submittals and approvals to prevent downstream bottlenecks.
- Identify bottlenecks early in the project

5. Facilitate Monitoring and Control

- Clear visibility of critical activities allows project managers to take corrective actions early.
- Variances between planned and actual dates highlight risks before they escalate.

6. Support Claims and Client Communication

- A well-maintained schedule provides evidence in case of disputes on delays, scope changes, or contractor responsibilities.

Golden Rule – Scheduling - *"If you don't control the schedule, the schedule will control you."*

5.6.4 Scheduling Techniques for HVAC Projects

A project schedule is only as effective as the technique used to build and monitor it. HVAC projects, with their multiple dependencies, benefit from a combination of simple visual tools and advanced analytical methods. There are a number of tools and software programs available today that can be used effectively to schedule tasks. Some of them are briefly described here.

I. Gantt Chart

- A bar chart that shows activities against time.
- Easy to create, widely understood by clients, consultants, and site teams.
- Helps visualise start and finish dates, overlaps, and sequencing.

A **Gantt chart** can show ducting, piping, and electrical cabling activities running in parallel in different zones of a building, with

Strength: Simple and intuitive. Most of the HVAC projects use Gantt Charts to show the progress of the project. Very easy to use.

Limitation: Does not highlight the critical path (the sequence of activities that drives project completion).

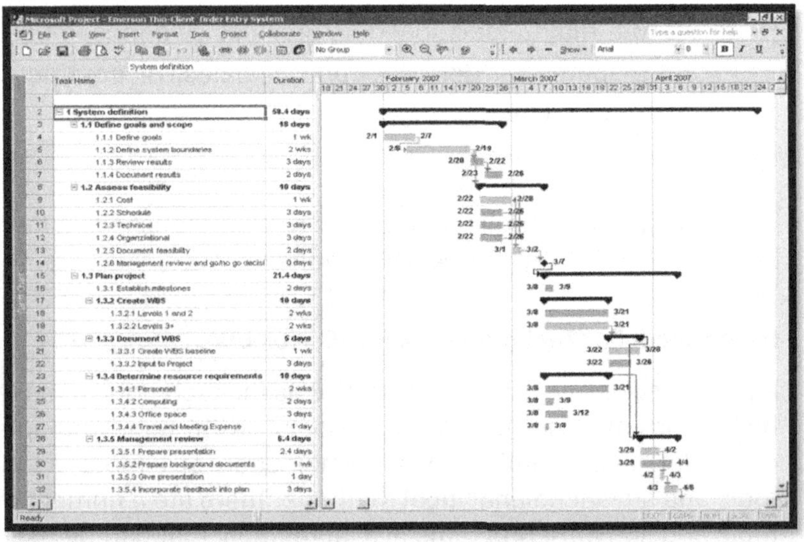

Example: Figure 5.3 GANTT Chart

II. Critical Path Method (CPM)

The critical path method is a widely used project scheduling method across many projects in the world. It gives a clear picture of the project and alerts the project manager to any probable delays much in advance.

CPM techniques

- Identifies the longest path through the project schedule (activities with zero float).
- Critical activities must be completed on time, or the entire project gets delayed.
- Non-critical activities have "float" and can be delayed within limits.
- Identifies project priorities
- Shows activity relationships and dependencies clearly
- Enables schedule analysis, helps in deciding crashing, fast tracking and resource levelling
- Reflects project strategy
- Promotes teamwork

Example:

In an HVAC project, activities like Pump Installation →chiller delivery → chiller installation → piping connection → testing & commissioning often form the critical path. If the chiller delivery slips, the entire project handover is delayed.

Strength: Provides clarity on must-do activities.

Limitation: Assumes activity durations are fixed and certain. Most non-critical activities become critical due to a lack of focus on them.

III. Program Evaluation & Review Technique (PERT)

One drawback with the CPM technique is, the time durations are approximate and at times highly padded up with safety. This may lead to a false sense of safety among the project teams.

PERT deals with uncertainty by using three-point estimates:

- Optimistic (O) time
- Most likely (M) time
- Pessimistic (P) time

Expected duration is calculated as (O + 4M + P) / 6.

PERT formula example: "VRF shipment time: O=4, M=6, P=10 weeks → Expected = (4 + 4×6 + 10)/6 = 6.33 weeks; Std Dev = (P–O)/6 = 1 week; use 1–2σ for buffer sizing."

Useful in projects with uncertain timelines (e.g., import clearances, client approvals).

Example:

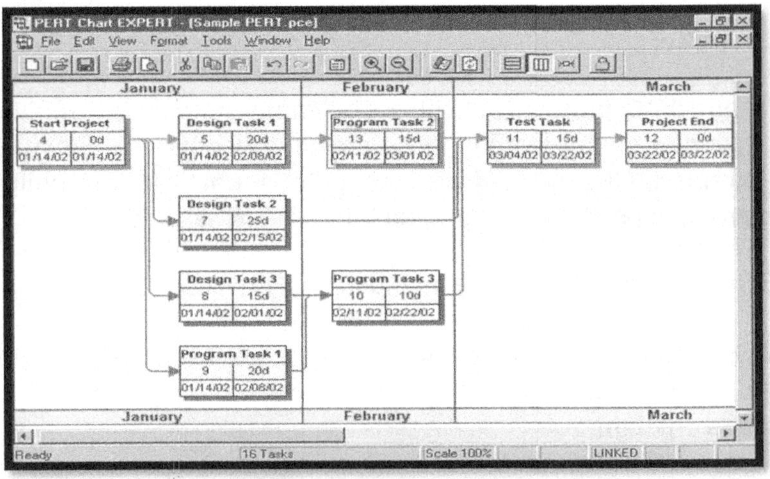

Example: Figure 5.4 PERT Chart

Delivery of VRF units may take 4 weeks (optimistic), 6 weeks (most likely), or 10 weeks (pessimistic). Using PERT helps account for this uncertainty in the schedule.

Strength: Incorporates risk into schedule planning.

Limitation: More complex to calculate and maintain. All line items may not have these kinds of three-point estimates.

IV. Critical Chain Project Management (CCPM)

Critical Chain Project Management (CCPM) builds on CPM but also considers resource constraints. It overcomes all the limitations of other techniques

- Uses buffers to protect the project schedule:
- Project Buffer (at the end of the project)
- Feeding Buffers (before critical chain activities)

Focuses on eliminating multitasking and ensuring resources are available when needed.

In a hospital HVAC project, the same skilled team may be needed for both ducting installation in the ICU and chiller piping work. CCPM ensures resources are not double-booked and uses buffers to absorb delays.

Strengths: Realistic in resource-limited environments (common in Indian projects).

Limitations: Requires cultural change — teams must focus on one task at a time. Not many are well-versed in the usage of this technique.

Example:

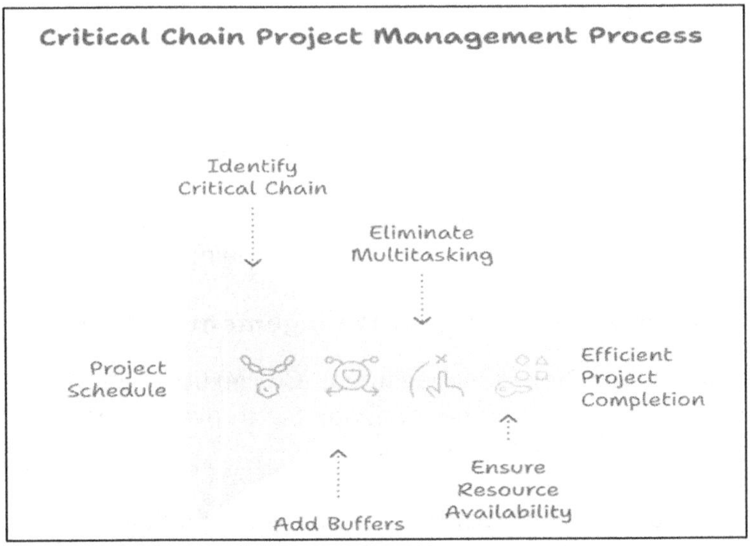

Example: Figure 5.5 Critical Chain Project Management Process

> **Golden Rule – Techniques:** *"Use Gantt for visibility, CPM for control, PERT for uncertainty, and CCPM for resource reality. Whatever tool you use, use it consistently to monitor and control."*

5.6.5 Developing a Baseline Schedule in HVAC Projects

Why baseline?

Schedule baselining is a critical step: it freezes the approved version of the project schedule at the start of execution so progress, delays and variances can be measured objectively. Without a baseline, monitoring becomes subjective, and claims/changes are hard to defend.

A good baseline makes every stakeholder aware of their timelines and the downstream impact of any delay.

The baseline schedule is the approved, frozen schedule that serves as the reference against which actual progress is measured. Project control — progress measurement, variance analysis and claims — is only meaningful after the schedule is baselined.

Link WBS → Activities → Dependencies → Schedule (Practical steps)

Follow these iterative steps to build a realistic baseline:

Step 1 — Break the WBS into manageable activities

Break deliverables into activities you can realistically monitor (for example: "Shop drawing submission", "Fabrication", "Delivery", "Installation", "Unit Testing").

Decide the level of detail pragmatically — too coarse and you miss control; too fine and nobody updates it.

Tip: If in doubt, include the activity. Every listed task or subtask should be monitored and updated.

Step 2 — Assign durations, resources and owners

For each activity, add: realistic duration (days), the resource(s) required (crew, crane, vendor), and a single accountable owner.
In the task comments, note tools required, standards to follow, and any special protocols.

Step 3 — Define dependencies

Define logical links (Finish-to-Start, Start-to-Start, etc.). Many schedules fail because dependencies are missed. Add Lags and Leads to the dependencies.

Examples:

"Submit technical submittals" → (then) "Consultant approval" → (then) "Release PO". The approval is a dependency for the PO.

"Duct installation" can start only after

a. duct fabrication drawings are approved and

b. The site is ready (wall openings, scaffold, power).

c. Duct installation can start two days before the fabricated ducts reach the site.

"Chiller delivery" should be preceded by "Chiller foundation ready" to avoid on-site waiting/demurrage. Lag will be at least 5-10 days for curing the frame

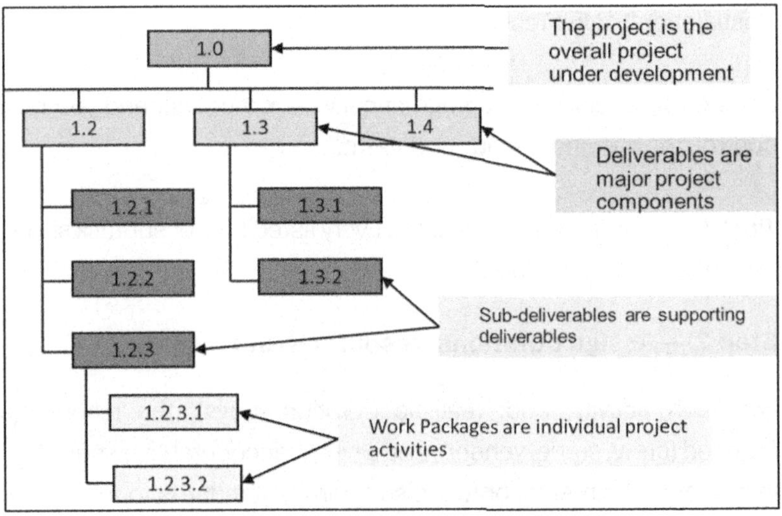

Figure 5.6 WBS in Layers

Step 4 — Include procurement milestones for long-lead items

Break procurement into sub-activities so the schedule shows the full chain and where delays can occur:

Select vendor → Technical submittal → Approval → Issue PO → FAT → Shipment → Customs clearance → Site delivery → Installation

Mark key checkpoints as milestones so they are visible at a glance (e.g., "Chiller FAT approved – Wk12", "Chiller site delivery – Wk18").

Step 5 — Incorporate identified risks

For risks captured during planning, add logic or contingency (lags, alternate vendors, buffer tasks) into the schedule where appropriate. Flag high-probability / high-impact tasks so the PM can watch buffer consumption.

Step 6 — Resolve resource conflicts

Review the schedule for resource clashes (same crew or crane assigned concurrently) and resolve by sequencing or resource levelling.

Iterate until conflicts are addressed. Then circulate the schedule to task owners for review and comments.

Step 7 — Review & Approve the Baseline

Hold a baseline review with PM, planning, procurement, QA/QC, key subcontractors, and client/consultant as required.

Freeze the schedule as Baseline v1.0 and document the sign-off (date, attendees).

Any future changes to the baseline must go through the Change Control Board (change request → impact analysis → approval → re-baseline if approved).

5.6.6 Do's & Don'ts for scheduling

Table 5.9 DOs and Don'ts of Scheduling

✅ Dos	❌ Don'ts
• Break down the scope into a **clear WBS → activities → dependencies** before building the schedule. **Baseline WBS** • Include **procurement milestones** (PO, FAT, shipment, customs, delivery) as real tasks. • Schedule **client/consultant approvals** explicitly — don't assume automatic clearance. • **Tie schedule to cost** load key tasks with estimated cost (PO value, manpower) for simple Earned Value tracking where needed. MS Project supports this. • Update progress **weekly**, capturing actual dates, % complete, and reasons for variance. • Use **buffers** (project/feeding) to absorb delays instead of padding every activity. • Train site engineers & supervisors to feed data into the schedule (mobile or Excel). • Make only one person accountable for the integrity of this project schedule	• Don't create **overly detailed schedules** that nobody updates (avoid "death by detail"). • Don't treat commissioning as **one line item** — break it into TAB, FPT, and seasonal tests. • Don't ignore **resource conflicts** (same crew booked for two tasks at once). • Don't rely on **verbal updates** — always document slippages and corrective actions. • Don't confuse **% complete** with progress — validate using **earned milestones**. • Don't ignore **seasonal constraints** (monsoon, holidays, power curtailments).

A Typical Schedule sample is given below with Predecessors

The sample schedule is given in the annexure

WBS	Activity	Owner	Dur (days)	Predecessor	Milestone?	Comments
2.4	Submit chiller technical submittal	Design	7	2	No	Include datasheets
2.3.1	Consultant approval – chiller submittal	Consultant	10	2.4	Yes	Approval triggers PO
3	Issue PO & vendor acknowledgement	Procurement	2	2.3.1	No	PO to include FAT clause
3.1	FAT at the vendor factory	QA/Proc	5	3	Yes	Vendor invites the site team
3.2	Shipment / Customs clearance	Procurement /Logistics	30	3.1	No	Include customs buffer
3.3	Site delivery & offload	Site Team	3	3.2	Yes	Check foundation readiness
4	Chiller installation & piping	Site Team	14	3.3	No	Requires crane & commissioning points
4.1	Pre-commissioning checks	QA/QC	7	4	No	Test points installed
4.2	TAB & FPT	Commissioning	10	4.1	Yes	Do not skip seasonal tests

Golden Rule: *"A baseline schedule is not just a plan — it is the contract between time and reality."*

5.6.7 Scheduling Tools for HVAC Projects

With technological advancements, a wide variety of tools are available today to schedule, monitor, and track projects effectively. Many ERP solutions even integrate procurement, inventory, and accounting modules with scheduling software to directly update progress.

The key is to select the right tool that:

- Is flexible and easy to use.
- Supports Gantt charts and dependency mapping.
- Provides auto-alerts (emails, WhatsApp groups, reminders).
- Is accessible from multiple devices (laptop, mobile, shared drives).
- Fits within the budget of the organisation.

Commonly Used Scheduling Tools

- **Excel Trackers**
  - Effective for small projects.
  - Simple S-curves and auto Gantt templates available.
  - Can be integrated with BOQs and linked to project documents.
  - Advanced users can automate tasks with VB scripts.

  - Limitation: becomes cumbersome for large/complex projects.

- **MS Project**
  - Widely used for medium to large projects.
  - Features: baselines, linking tasks, resource allocation, critical path tracking, dependency mapping, assigning costs, and earned value management.
  - Cloud version allows mobile updates by field staff.
  - Limitations: Requires training; license cost may be high for small firms.

- **Primavera / Project Libre / Standalone Apps**
  - Preferred for very large, multi-stakeholder projects.
  - Offers powerful resource levelling, risk analysis, and integration with enterprise systems.
  - Limitations: Steep learning curve; not practical for small HVAC contractors.

5.6.8 Updating & Controlling Schedule

- Weekly/monthly updates.
- Progress % tracking.
- Recovery planning in case of delays.

5.6.9 Practical Scheduling Challenges in Indian HVAC Projects

- Dependency on civil/electrical readiness.
- Delayed approvals/drawings.
- Import logistics & customs delays.
- Limited trained resources.

Case Study – Effective Scheduling for Timely Handover
Project: Corporate IT Campus – Hyderabad
Scope: Central chilled-water plant (3 × 500 TR chillers), VRF systems for special areas, air distribution for 6 office blocks.

- **Scenario:**
  The contractor prepared a baseline schedule that integrated civil readiness, long-lead procurement, and commissioning milestones right from the planning stage.
- Long-lead equipment like chillers and cooling towers was tracked with detailed sub-activities (submittal, FAT, shipment, customs clearance, site delivery).
- Dependencies such as "ceiling closure only after duct pressure test" and "piping insulation only after hydro test clearance" were explicitly modelled.

- Weekly progress meetings included a schedule variance dashboard with milestones, ensuring accountability of each activity owner.

- When customs clearance for VRF units was delayed by 10 days, the project team immediately re-sequenced ducting and electrical works in parallel, using available manpower effectively.

Outcome:

- Handover achieved two weeks ahead of the client's target.

- Zero claims for delay — client released retention money without dispute.

- Contractor avoided idle manpower cost and gained a repeat order from the same client.

Lesson Learned:

- Proactive scheduling with clear dependencies and continuous monitoring transforms potential delays into opportunities to re-sequence work, ensuring timely delivery and client satisfaction.

Golden Rule – Scheduling in Practice: *"A live schedule is a shield — it protects the project from surprises by turning risks into manageable actions."*

5.7.10 Conclusion

Scheduling is the backbone of project execution. In HVAC projects, where dependencies are high and delays in one activity ripple through multiple trades, a well-prepared and well-monitored schedule separates successful projects from chaotic ones.

A schedule is not just a list of dates — it is a management tool that integrates work breakdown structure, procurement, installation, and commissioning milestones into a single roadmap.

By baselining the schedule, project managers gain an objective reference point to measure progress, highlight risks, and justify claims. More importantly, a disciplined scheduling process ensures accountability: every activity has an owner, every dependency is mapped, and every milestone is visible.

> **Golden Rule**: *"A schedule without monitoring is a wish list. A baseline schedule with active control is the project manager's most powerful weapon."*

5.7 Project Cost and Budget Management in HVAC

"You can control scope, you can control schedule, but if you don't control cost, you don't control the project."

5.7.1 Introduction

In HVAC projects, cost is the most visible performance parameter — often the deciding factor for both contractors and clients. A project delivered on time but over budget is rarely considered successful. Cost management in HVAC goes beyond estimating BOQs; it involves forecasting cash flows, monitoring expenses, managing variations, and protecting margins against risks.

In the Indian context, where HVAC contracts often have thin profit margins (5–10%), delayed payments, scope creep, or unexpected claims can quickly erode profitability. Cost management is, therefore, not just a financial activity; it is a survival skill for HVAC contractors.

Golden Rule – Cost Control: *"Revenue is vanity, profit is sanity, but cash flow is the reality."*

5.7.2 Process flow

The Contract BOQ (Bill of Quantities) forms the starting point for project estimation and budgeting. In a typical HVAC project, **nearly 70% of the total cost comes from materials** — ducting, insulation, piping, electrical cabling, valves, and air distribution products. However, this dependency on material quantities makes cost control extremely challenging. **Why? Because quantities are rarely static.**

➢ Once drawings are baselined, quantity take-offs are compared against BOQ items and deviations are approved. A cost baseline (budget) is established.

➢ In an ideal scenario, if drawings and site conditions never changed, the required materials could be ordered precisely as per approved quantities.

➢ In real scenarios, HVAC projects face frequent changes:

- Duct/pipe routings get altered on site.
- Additional conditioned areas are added mid-project.
- Materials are wasted, damaged, or rusted before use.

  - Excess materials are delivered but no longer usable after design changes.

- **The result:** significant cost variations between estimated, procured, installed, and measured quantities.

Another recurring problem in India is work carried out through oral instructions, bypassing formal change orders. This leads to disputes, unaccounted costs, and strained client–contractor relationships.

The cost management process is briefly explained here.

Table 5.10 Project Cost Planning Process flow

Inputs	Process	Outputs
o Contract documents (BOQ, tender specs, payment terms) o Approved shop drawings & technical submittals o Procurement plans & vendor quotes o Risk register o Previous project cost data	o Cost Estimating (materials, labour, subcontractors, overheads) o Cost Budgeting → Develop Cost Baseline (time-phased budget linked with schedule) o Cash Flow Forecasting o Cost Control → Track actuals vs. baseline; Manage variations/claims o Earned Value Management (optional for large projects)	o Project Cost Estimate o Cost Baseline o Cash Flow Projection o Variance reports o Forecasted cost to complete

Case Study – Cost Overrun Due to Poor Planning

Project: HVAC installation for a shopping mall, Pune (₹8 crore contract).

- Contractor underestimated ducting quantities during bidding (used tender BOQ without site validation).
- Imported AHUs were delayed at customs, leading to airfreight charges of ₹20 lakhs.
- Consultant insisted on additional sensors and control wiring not included in the scope.
- Client's billing cycle required approvals from 3 agencies, stretching payments to 90 days.

Result:

- Actual cost exceeded baseline by ₹1.2 crore (15%).
- Contractor faced severe cash flow issues and borrowed working capital at high interest.
- Profit margin wiped out; dispute with client dragged into arbitration.

Lesson Learned: Cost management must start from contract review + validated BOQ, include realistic contingencies, and ensure cash flow is aligned with payment terms.

5.7.3 Best Practices in Cost Management (HVAC)

1. **Link Cost to Schedule:** Every cost element (ducting, piping, chiller procurement) must be tied to schedule milestones.

2. **Maintain a Cost Baseline:** Lock an approved cost baseline against which actuals are tracked.

3. **Cash Flow Forecasting**: Always prepare monthly inflow vs. outflow forecasts — identify gaps early.

4. **Variation Management:** Track scope changes through a Variation Register and seek written approval before execution.

5. **Vendor Management:** Negotiate payment terms with suppliers to align with client billing cycles. Get the right prices and other terms such as warranty period, Training, tools and other such requirements as per the Customer contract.

6. **Vendor Management:** Negotiate payment terms with suppliers to align with client billing cycles. Get the right prices and other terms such as warranty period, Training, tools and other such requirements as per the Customer contract.

7. **Historical Data:** Use past project data to validate estimates and contingencies.

8. **Regular Reviews:** Conduct monthly cost reviews with the project team and management.

Table 5.11 Cost Variation Tracker

WBS/Activity	Planned Cost	Baseline Start	Baseline Finish	Actual Cost (Lakhs)	Variance (Lakhs)	Remarks
2.1 Ducting	120	01-Jul-25	31-Aug-25	130	10	Extra GI Sheets
2.2 Piping	95	10-Jul-25	25-Aug-25	100	5	Extra welding joints
3.1 Chiller Procurement	250	15-Jul-25	15-Sep-25	260	10	Customs clearance delays
3.2 AHU Procurement	150	20-Jul-25	20-Sep-25	145	-5	Discount from vendor
4.1 Installation Labour	80	01-Aug-25	15-Oct-25	70	-10	Less manpower used

5.7.4 Cash Flow Reality: Payment Terms & Credit Periods

In HVAC projects, even when costs are well-estimated on paper, cash flow gaps often derail execution. The challenge lies in how payments, credit periods, and quantity variations interact in real life:

1. **Payment Terms Are Uneven:** Every client has their own billing procedures and approval cycles. Contractors must procure materials upfront from multiple "approved suppliers," but the billing and certification process with the client can take weeks or even months. This mismatch between inflow and outflow creates a serious cash strain.

2. **Supplier Credit Periods Add Pressure:** Vendors demand payments within fixed credit periods (30–45 days in most cases), while client disbursements often come much later. Contractors end up bridging the gap, and unless margins

are carefully managed, working capital erodes rapidly. In some cases, vendor payments also get delayed, thus losing credibility in the market.

3. **Delayed Measurements & Billing:** Site measurements, consultant verifications, and certification delay often stall invoicing. Even when the work is physically completed, billing gets stuck in paperwork, leaving contractors to fund ongoing site progress out of pocket.

4. **Excess Quantities Beyond BOQ:** In practice, 70–80% of projects see extra quantities supplied on-site due to revised routings, added areas, or design modifications. Yet, payments for these variations are released only after a formal amendment is issued — a process that may take months. Meanwhile, contractors are forced to supply, install, and carry the cost burden without reimbursement.

Bottom Line: Unless contractors actively manage payment terms, align supplier credit cycles, and negotiate provisional approvals for variations, cash flow gaps can wipe out margins even in otherwise well-managed projects.

Golden Rule – Payment Terms: *"Cash flow is not about profit or loss on paper — it is about timing. If inflows lag behind outflows, even a profitable project can sink."*

Case Study – When Cash Flow Kills a Project

Project: HVAC Package for a Commercial Mall, Pune (₹15 crore contract value)

Scenario:

The contractor started execution confidently. BOQ was approved, and materials were ordered as per the site requirements. However, several site-level issues created a cash flow nightmare:

- **Delayed Measurements:** Consultant approvals for monthly running bills took 6–8 weeks instead of the contractually agreed 3 weeks.
- **Payment Terms Mismatch:** Contractor had to pay vendors within 30 days, but clients released payments only after 90+ days.
- **Excess Quantities Beyond BOQ:** Additional ducts and insulation (due to design changes) were installed as per oral site instructions, but no formal amendment was issued for 6 months.
- **Blocked Working Capital:** Imported AHUs and chillers worth ₹3 crore arrived, but the site was not ready. They lay idle at the site, payments stuck, while the contractor had already paid advances to suppliers.
- **Credit Overstretch:** To maintain progress, the contractor borrowed heavily to pay vendors. Interest costs further eroded margins.

Outcome:

By mid-project, the contractor's cash reserves were exhausted. Vendor payments were delayed, site manpower was reduced, and progress slowed. The client levied Liquidated Damages (LDs) for delay. Eventually, the contractor completed the project but made almost zero profit — in fact, lost money due to penalties, idle inventory, and interest costs.

Lesson Learned:
Technical execution was sound, but poor cost management and weak cash flow planning turned a promising project into a financial failure.

> **Golden Rule:** *"Profit is not lost in execution — it is lost in poor cash flow control."*

5.7.5 Best Practices for Cash Flow Management in HVAC Projects

- Align payment terms with supplier billing cycles during contract negotiation.
- Don't underestimate the impact of delayed measurements; build billing schedules around realistic site measurement dates.
- Don't allow project cash flow to run on assumptions — base it on real approved quantities and certified bills.
- Maintain a cash flow register: inflow vs outflow linked to project milestones.
- Keep a buffer for delayed approvals and retentions (5–10%).
- Get the deviation statement approved soon after the take-off quantities are arrived at.
- Don't rely on oral instructions for additional work; always document with written site instructions/variation orders.
- Arrange to return excess materials from the site immediately.

- Don't overlook contract clauses on material return policies — some contracts do not allow material returns until final handover, creating a cash burden if items are sent prematurely.

- Don't dispatch excess materials to the site without formal approval — it locks cash and adds financial strain.

- Submit deviation statements early and get amendments approved before supplying beyond the BOQ.

- Track certification timelines for bills (consultant → client → finance) and factor delays into cash flow forecasts.

- Don't ignore credit periods - mismatched supplier/ client cycles erode margins.

Golden Rule – Cash Flow Awareness: *"Train your site supervisors and engineers on cash flow management. Technical execution builds systems, but financial discipline sustains projects.*

5.8.6 Do's & Don't s – Project Cost Management

Table 5.12 Do's & Don'ts – Project Cost Management

✅ Dos	❌ Don'ts
○ Do prepare a detailed cost baseline and cash flow before mobilisation. ○ Do align subcontractor/vendor payments with client billing. ○ Do maintain a Variation Register and escalate early ○ Do use checklists to monitor material wastage. ○ Do review cost status monthly with PM + Accounts team.	○ Don't depend solely on tender BOQ — validate quantities at the site. ○ Don't execute additional work without written approval. ○ Don't ignore delays in billing approval cycles. ○ Don't mix overheads and direct costs — track separately. ○ Don't wait till the project ends to reconcile actual vs. planned cost

5.6.7 Conclusion

Cost management is not just about saving money — it is about ensuring financial discipline, protecting margins, and sustaining business credibility. In HVAC projects, where technical excellence must go hand-in-hand with commercial acumen, a project manager's ability to control costs often determines the success or failure of the contract.

A well-managed cost plan achieves three goals:

- **Protects the profitability** of the contractor.
- **Builds trust** with the client by avoiding disputes.
- **Ensures cash flow stability** to keep the project moving.

Golden Rule - HVAC Cost Management: *"Projects don't fail because costs were high — they fail because costs were uncontrolled."*

5.8 Resource Management in HVAC Projects

A project is only as strong as the people who execute it — tools and schedules mean little without the right resources in place."

5.8.1 Introduction

Resource management is the art of ensuring the **right people, with the right skills, are available at the right time**. In HVAC projects, resource challenges are often the biggest cause of delays, cost overruns, and quality issues. A perfectly planned schedule can collapse if duct erectors are not mobilised on time or if TAB engineers are missing during commissioning. Effective resource management ensures smooth progress from design through to handover.

In the HVAC industry, it is common to subcontract work such as ducting, piping, insulation, and electrical installations. These subcontractors need close supervision — not just assigning work, but ensuring execution is aligned with approved method statements, standards, and the Owner's Project Requirements (OPR). Supervision itself demands engineering judgment: random inspections on the job, corrective instructions when needed, and technical guidance to resolve site conflicts.

Getting the **right supervisors and engineers**, with the ability to read shop drawings, apply engineering knowledge, and communicate effectively, is therefore critical. Yet, many engineers step into project management roles based purely on technical experience, without exposure to formal project management practices. This skill gap often puts projects at serious risk.

That is why it is essential to **train resources** not only in technical aspects but also in commercial and soft skills — from communication and negotiation to planning and documentation. This book is designed to serve as a practical guide for HVAC professionals stepping into project management roles. In this chapter, we will explore the process of resource management and how it can become a strategic driver of project success.

➢ **In India, the major causes of HVAC project delays are resource shortages, poor subcontractor management, and unbalanced manpower deployment.**

5.8.2 Resource Planning Process Flow

Table 5.15 Resource Management Process Flow

Inputs	Process	Outputs
• Project BOQ and Drawings • Baseline schedule • Procurement schedule • Site readiness & constraints • Subcontractor agreements	• Prepare manpower histogram (by skill: duct fitter, welder, insulation team, electricians). • Map resources against activities (resource loading). • Align vendor delivery schedules with site manpower availability. • Identify peak periods and clashes with other trades.	• Resource Deployment Plan (by week/zone). • Resource Histogram / S-Curve. • Vendor Mobilisation plan • Resource levelling • Resource Responsibility Matrix

Case Study – Resource Bottleneck

Project: Commercial Mall – Pune

Issue: The PM scheduled ducting, piping, and insulation to progress in parallel, but failed to check whether subcontractors had enough manpower. The ducting subcontractor deployed 80 fitters, while piping had only 15 welders instead of the planned 40. As a result, ceilings could not be closed, delaying interiors by 6 weeks.

Lesson Learned: Schedules without resource validation are meaningless. Always align subcontractor commitments with the project plan.

5.8.3 Roles & Responsibilities: RACI, Approval & Escalation Matrices

One of the most common issues in HVAC projects is confusion about **"Who is responsible?"**

When the roles and responsibilities are not well defined, it becomes everyone's responsibility. A delayed approval or an unresolved site issue can derail weeks of good work.

The following documents are generally prepared to manage resources.

- **RACI Matrix** defines who is ***Responsible, Accountable, Consulted, and Informed*** for each task. This matrix clearly defines who is responsible for which activity.
- For example, the *site engineer is responsible for duct installation, the project manager is accountable, the consultant is consulted, and the client is informed.*
- **Approval Matrix** specifies authority for approvals. This is very important for all technical documents, measurements and submission of bills for payments.
- e.g., material submittals by QA/QC → reviewed by consultant → approved by client.
- Joint Measurement sheets → reviewed by consultant → approved by client.

- **Escalation Matrix** sets clear pathways when issues arise. An issue log is maintained by the PMO/ Project manager. It is very important to resolve any issues identified by any team member in the project. If the issue is not getting resolved within a reasonable period, then it should be escalated to the next higher authority to resolve it without any further delay.

Table 5.16 RACI Matrix

Activity	Project Manager	Planning Engineer	Site Engineer	QA/QC Engineer	Sub-contractor	Client/ Consultant
Prepare Resource Plan	A	R	C	C	C	I
Validate Subcontractor availability	C	C	A	C	R	I
Manpower Deployment on Site	C	A	R	A	R	I
Track Resource Utilisation	A	R	R	C	C	I
Align Vendor Deliveries	R	C	A	C	C	I
Weekly Resource Review Meeting	A	R	C	C	C	I

An escalation matrix is prepared in the planning phase of the project itself for different categories of issues as described below.

- **Technical Issues**: Any issue related to technical aspects of the projects will be solved by the contractor's technical team. If not resolved, escalate to the project manager → consultant. Issues related to technical submittals, design issues, installation conflicts, duct rerouting, location of

Cooling Tower, Chiller or Air-cooled ODUs and any other issues are identified and resolved by this team

- **Commercial/ Contract issues** → escalate to contracts manager →Project Manager →client finance. Measurements, Payment terms, Payment process, Amendments to Quantities, BOQ Quantity variations, price escalations, new items outside BOQ, Non-standard items, rework items pricing, etc.

- **Site coordination Issues** → escalate to contracts manager →Project Manager → escalate to planning head. →Escalate to a consultant

- **HSE compliance issues**: escalate to the contracts manager →Project Manager →Safety Head of the projects. Health, Safety and Environment protocols are to be followed strictly at the site. Any deviation will pause the progress of the work. An escalation matrix is created to ensure that these incidents are escalated appropriately and resolved.

- Without these matrices, problems often linger, creating frustration and wasted time.

> **Golden Rule:** *"Escalation is not on any person. It is on issues and processes. Fix the process, not the blame."*

5.8.4. Team Requirements in HVAC Projects

Every phase of the HVAC life cycle requires Specialised expertise. Having the right mix of design, execution, commercial, compliance, and subcontracting teams is critical for success.

- **Design Phase**: HVAC designers, draftsmen, BIM coordinators.
- **Execution Phase**: Project manager, planning engineer, site engineers, and supervisors.
- **Testing & Commissioning Phase**: TAB engineers, QA/QC inspectors.
- **Commercial Functions**: Procurement team, billing team, and finance.
- **Compliance**: Safety officers, documentation managers.
- **Subcontractors**: Duct fabricators, piping contractors, insulation crews, electricians, BMS integrators.

Table 5. 17 Sample Escalation Matrix

Category of Issue	Level 1 (Immediate Owner)	Level 2 (Escalation)	Level 3 (Final Escalation)	Resolution Time	Examples
Technical Issues	Contractor's Technical Team	Project Manager	Consultant / Client Technical Authority	48 hrs	Technical submittals, design conflicts, duct rerouting, chiller/cooling tower location
Commercial / Contract Issues	Contracts Manager	Project Manager	Client Finance / Contract Authority	7 days	BOQ variations, pricing disputes, new items, non-standard materials, rework pricing
Site Coordination Issues	Site Engineer / Supervisor	Project Manager	Planning Head / Consultant	48 hrs	Site readiness, manpower shortages, work sequence conflicts
HSE Compliance Issues	Site HSE Officer	Project Manager	Project Safety Head	Immediate / 24 hrs	PPE non-compliance, unsafe scaffolding, unsafe lifting practices

Planning resources in advance is a top priority. The project manager, together with the team, identifies requirements at each stage. In the early project phase, the team may consist only of the design staff. As the project progresses, procurement specialists, supply chain coordinators, site supervisors, and contractors get mobilised. The timing of each resource's start date and duration of deployment is critical.

Example: A project may stall if duct erection is completed but insulation subcontractors are unavailable — delaying ceiling closure and subsequent works.

5.8.5. Project Allocation and Resource Levelling

Resource allocation ensures people are neither idle nor overloaded. Proper allocation keeps productivity high and avoids costly downtime.

- **Manpower histogram**: Shows resource peaks (e.g., maximum welders required during piping peak) and troughs.
- **Resource smoothing**: Balances workloads, avoids sudden bottlenecks.
- **Subcontract allocation**: Specialised tasks (e.g., copper brazing, AHU foundation works) are allocated to skilled teams only

5.8.6. Assessing Training Needs

A recurring cause of poor quality in HVAC projects is **skill gaps**. Getting resources with the right capabilities is always a challenge. The resource manager, along with domain heads, should assess each team member's competencies and identify training needs **before allocation**.

- **Safety Training**: PPE usage, working at heights, confined space permits.
- **Technical Training**: Brazing techniques, welding, duct sealing, and insulation practices.
- **Soft Skills**: Communication, teamwork, documentation.
- **Induction for Subcontractors**: Orientation on project rules, quality standards, and safety norms.

Training records must be maintained and verified on site.

5.8.7. Performance Measurement & Appreciation

Resources perform best when monitored and recognised:

- **Metrics:**
  - Productivity (e.g., ducting m^2/day, piping m/day).
  - Quality (NCR rates)., Zero Defects, CAPA
  - Safety compliance.
- **Recognition:**
  - "Team of the Month" certificates.
  - On-site appreciation boards.
  - Linking strong performance to future project opportunities.

Example: A ducting team consistently exceeded daily targets with zero rework → recognised with appreciation letters → boosted morale across trades.

5.8.8 Project Release & Lessons Learned

At project completion, resources need a structured release:

- **Project Release:** Smooth demobilisation of site teams, proper closure of subcontract agreements.
- **Lessons Learned:** Documenting challenges (e.g., "TAB team mobilised late caused handover delay") and improvements ("BIM clash detection reduced duct rerouting").

These insights should feed into the **organisation's knowledge base**, ensuring future projects don't repeat the same mistakes.
Best Practices in Resource Planning

- Link manpower plans directly to schedule milestones.
- Prepare **weekly resource deployment charts** for each trade.
- Conduct joint planning meetings with subcontractors/ vendors to verify availability.
- Track **actual vs planned** manpower through daily reports.
- Use simple Excel-based S-curves to visualise manpower loading and cash flow.
- Crosstrain teams (e.g., duct fitters on insulation) for flexibility during crunch times.

Practical Tip

- *Subcontractors may bring manpower, but it is the **project team's responsibility** to provide clarity, supervision, and standards. Never assume that subcontractors will deliver quality without guidance. Clear instructions, method statements, and continuous oversight are the backbone of successful execution.*

Table 5.14 Do's & Don'ts – Project Resource Management

✅ Dos	❌ Don'ts
○ Do prepare a resource histogram linked to the WBS and schedule. ○ Do validate subcontractor's resource commitments in writing. ○ Do plan for peak manpower in advance (scaffold, cranes, safety supervisors). ○ Do consider seasonal variations (festivals, monsoon, labour migration). ○ Do review resource plans weekly in site coordination meetings.	○ Don't assume vendor promises without backup documents. ○ Don't overload the site with idle manpower — it increases costs without progress. ○ Don't neglect site facilities (toilets, storage, access) needed to support resources. ○ Don't forget to align the resource plan with cash flow — unpaid subcontractors will demobilise suddenly. ○ Don't leave critical equipment (cranes, welding sets, TAB instruments) to last-minute booking.

5.8.9 Conclusion

Resource management is more than simply assigning names to tasks. It is about building a **cohesive ecosystem of responsibilities, approvals, skills, and support**. In HVAC projects, where coordination across trades is critical, resource gaps can cause cascading delays and rework. A well-planned resource strategy ensures that the right people are mobilised, trained, monitored, appreciated, and released at the right time. By formalising roles through RACI and escalation matrices, balancing manpower through histograms, and investing in training and recognition, HVAC project managers can transform resource planning into a driver of project success rather than a bottleneck.

Golden Rule: *"Even the best schedule fails without the right resources at the right time."*

Table 5.15 Chapter Summary Sheet

Aspect	***Key Point***
Purpose	*Ensure manpower, tools, and subcontractors are aligned with the schedule.*
Inputs	*BOQ, Drawings, Schedule, Vendor delivery plan, Subcontractor agreements.*
Process	*Resource loading → Deployment plan → Weekly tracking.*
Outputs	*Resource histogram, Manpower S-curve, Vendor mobilisation plan.*
Best Practices	*Weekly manpower charts, subcontractor review meetings, and cross-training.*
Pitfalls	*Idle labour, vendor delays, missing cranes/tools, and ignoring festivals/holidays.*
Golden Rule	*"Plan resources as carefully as activities — both must move together."*

5.9 Project Procurement Management in HVAC

The bitterness of poor quality remains long after the sweetness of low price is forgotten." – Benjamin Franklin

5.9.1 Why Procurement Planning Matters in HVAC Projects

In HVAC projects, procurement plays a central role in ensuring smooth execution. Typically, **70% of the project cost is materials, 20% is installation/labour**, and the remaining **10% covers testing, commissioning, and other costs**. With such a large material component, the success of the project hinges on getting the right materials, at the right time, and within budget.

However, procurement is not just about placing purchase orders — it is a fine balancing act:

Ordering Too Early:

- Leads to high inventory carrying costs.
- Requires additional handling, lifting, and storage space.
- Creates risks of theft, damage, or deterioration at the site.

Ordering Too Late:

- Causes material shortages.
- Delays downstream installation activities.
- Triggers re-sequencing of work and costly idle labour.

Order Quantity:

- Order quantity is more than BOQ; wastage and cash flow are impacted.

- Order Quantity is low, transportation cost increases, and a shortage of materials.

Thus, procurement must be treated as a systematic process:

For effective procurement planning, the following skills are required

- **Technical Knowledge:** This is very important for buying materials for HVAC projects, as most of the materials are designed and built to order.
- **Planning Skills**: Align material deliveries with the construction schedule.
- **Negotiation Skills:** Secure the best terms for pricing, payment, and delivery.
- **Collaboration Skills**: Coordinate between vendors, consultants, the site team, and logistics.

When executed well, procurement becomes **the silent enabler** of smooth project execution, avoiding both the chaos of material shortages and the hidden costs of overstocking.

> **Golden Rule – Procurement Planning:** *"Procurement is not about buying early or late — it is about buying smart, in sync with the project. "*

Case Study: Commercial Complex HVAC Project – Procurement Delays and Cost Overruns

A contractor was awarded HVAC works for a large commercial complex in Hyderabad. The scope included central chilled water systems with AHUs, chillers, pumps, ducting, insulation, and BMS controls. The project was scheduled for completion in 18 **months.**

What Went Wrong?

1. **Delayed Procurement of Long-Lead Items**

- The chiller submittals were raised late, and the PO was delayed by almost 3 months.
- By the time the chillers were delivered, the civil foundation had been ready for months — manpower idled, costing the contractor heavily.

2. **Poor Planning of Bulk Materials**

- Insulation materials were ordered in one huge lot without considering storage space. Sheets were stacked at the site in open areas, leading to rusting, rejections, and re-fabrication.
- All pipes were delivered at the site initially without ascertaining the readiness of the site. The client refused to make the payment for the material as it was not in the budgeted billing cycle. This resulted in material lying at the site, and later, some of the routings changed. This leads to a lot of pipe sizes being redundant at the site.

3. **Vendor Payment Delays**

- Pumps and AHUs faced delivery delays as vendors withheld dispatches until payments were cleared.
- This further pushed back site installation.

4. **Losses Due to Security Issues**

Though many project managers may not admit in public, it is a regular phenomenon at sites that run for more than 6 months.

- Copper piping and insulation materials were delivered months before they were needed.
- Lack of secured storage led to theft and wastage, forcing repeat orders at higher prices.

5. Rework Due to Outdated Drawings

- Ducting and piping fabricated early were based on outdated drawings.
- When updated drawings were released, large sections had to be dismantled and redone. Fabricated ducts were to be scrapped.

The Net Impact

- Project completion delayed by 8 months
- Project cost almost doubled due to:
    - Idle labour costs.
    - Rework on ducting and piping.
    - Repeat procurement for stolen/damaged materials.
    - Heavy penalties from the client.
- Contractor's reputation suffered, resulting in loss of future projects.

Lessons Learned

- Procurement must be synchronised with the project schedule — not rushed and not delayed.

- Secure storage and inventory control are critical for high-value items like copper piping.
- Payment terms and vendor management must be handled proactively to avoid dispatch delays.
- Only fabricate/install with approved drawings to avoid expensive rework.

> **Golden Rule – Procurement Team:** *"A procurement team rarely gets noticed when deliveries are on time — but the entire project feels the impact when they go wrong."*

Every rupee lost in procurement planning multiplies downstream — in delays, rework, and penalties.

As these are not manufactured in bulk, and only a few approved vendors are allowed to supply these materials, negotiating with the vendors for better pricing and better commercial terms is always a challenging activity. Many projects incur losses due to a lack of knowledge or delayed procurement.

5.9.2 Procurement Scope in HVAC Projects

In HVAC projects, procurement is generally divided into **high-side** and **low-side** equipment.

- **High Side Equipment** includes the major refrigerant and mechanical systems that form the backbone of HVAC: chillers, cooling towers, VRF systems, packaged units, pumps (primary, secondary, condenser water), control panels for pumps, air handling units (AHUs), and treated fresh air units (TFAUs).
- **Low Side Equipment and Materials** consist of all the associated works required to complete the air conditioning system: fan coil units (FCUs), ducting, air

distribution system, air distribution products and terminals, fire dampers, volume control dampers, piping and valves, insulation, filtration systems, air separators, expansion tanks, electrical panels, cables, earthing, and building management system (BMS) components such as sensors, energy meters, VFDs, pneumatic/electronic/DDC controls.

5.9.3 Why is HVAC Procurement Complex?

Unlike standard construction materials, HVAC equipment is mostly **engineered to order**. The same item (e.g., a chilled water pump) can have multiple variables depending on the project's specific design intent:

- Application (chilled water, condenser water, hot water)
- Type of pump (end suction, vertical inline, split case)
- Hydraulic parameters (flow rate, head)
- Materials of construction (body, impeller)
- Efficiency grade (meeting ECBC, IGBC, or consultant standards)
- Coupling type (direct, flexible)
- Motor specifications (efficiency, stages, voltage)

Because of this, most high-side equipment is **not manufactured in bulk** but produced **only against an approved data sheet**. Before procurement, detailed **technical submittals** — including drawings, performance sheets, and compliance statements — must be reviewed and approved by the consultant or project manager.

5.9.4 Challenges in HVAC Procurement

- **Limited vendor base** – Only a few manufacturers are approved for Specialised items.
- **Long manufacturing lead times** – Equipment may take weeks/ months after approval before dispatch.
- **Commercial negotiations** – Pricing and payment terms are tough to optimise when vendor options are limited.
- **Knowledge gaps** – Poor understanding of specifications or late procurement often leads to costly project delays and penalties.
- **Storage and Reusability** As most of these units are procured on specific project requirements and drawings, any excess materials will be a burden on the contractor. These materials may not be reusable in other projects due to the specifications and inspection before delivery clause. Even commonly used items such as Pipes and insulation may also not be reusable in other projects. Once any excess material returns from the project, the condition of the material is to be inspected before being taken back into the inventory system.

Thus, HVAC procurement requires not only technical knowledge of equipment and systems but also **project management skills in planning, vendor management, and negotiations**.

> **Golden Rule – Procurement Scope:** *"Every HVAC project is unique — equipment is engineered to order, not off the shelf. Treat procurement as engineering, not just purchasing."*

5.9.5 Procurement Management Process in HVAC

Every HVAC project has its own procurement process, which may vary slightly across organisations. Companies with ***ISO certification*** usually have well-defined procurement procedures and vendor qualification policies covering technical, commercial, and supply capabilities. Regardless of organisational differences, the essence *of HVAC procurement* follows a systematic approach.

Step 1 – Contract & Scope Review

- Read the contract thoroughly for technical specifications, delivery instructions, testing, performance, and installation requirements.
- Review BOQ and drawings for procurement needs.
- *Ensure drawings are approved before proceeding.*

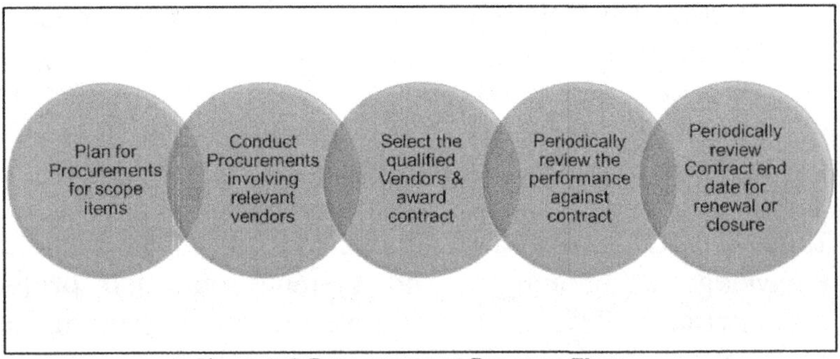

Figure 5.7 Procurement Process Flow

Step 2 – Identify Long-Lead and Critical Items

- Identify long-lead equipment (e.g., chillers, cooling towers, AHUs, pumps, panels).

- Confirm final quantity take-offs zone-wise from approved drawings.
- Map long leads to the critical path/ Critical chain of the project schedule.

Step 3 – Align Procurement with Project Schedule

- Revise project schedule to include a procurement schedule linked to milestones.
- Get approval from the competent authority and relevant stakeholders.

Step 4 – Vendor Shortlisting and Inquiry

- Prioritise long-lead items and immediate site needs.
- Float inquiries to qualified vendors.
- Finalise shortlist based on capability and compliance.

Step 5 – Technical Submittals

- Prepare and submit complete submittals with catalogues, calculation sheets, compliance statements, and references to standards.
- Tip: Incomplete submittals are a major reason for delays; don't blame consultants later.
- Track submittal status (submitted, under review, approved).

Step 6 – Negotiation and Purchase Orders (POs)

- Prepare comparison sheets of vendors (technical parameters, delivery schedules, quality, commercials).
- Negotiate terms and issue POs.

- Define clearly:
    - Supplier's scope vs non-scope items.
    - Payment terms (advance, milestone-based).
    - Sub-assemblies (ancillary items may need separate POs).

Step 7 – Vendor Progress and Inspections

- Obtain PO acknowledgement and acceptance.
- Follow up on manufacturing and delivery schedules.
- Arrange Factory Acceptance Tests (FATs) if required.
- Fulfil all commercial formalities before shipment.

Step 8 – Logistics and Delivery

- Logistics costs are significant and often underestimated.
- Coordinate procurement with logistics, inventory, and site teams.

  Plan for:
    - Transport mode (full/half truck load, special trailers).
    - Site access restrictions and crane/forklift requirements.
    - Transit insurance, safety protocols.
    - For imports: engage reliable customs clearance agencies and plan for buffer time.

Step 9 – Material Receipt and Quality Checks

- Ensure cranes/forklifts are ready for unloading.
- Conduct QA checks on receipt of materials.
- Maintain proper inventory registers and secure storage.
- Protect materials with lighting, ventilation, and theft prevention measures.

> **Key Takeaway:** *Procurement is not just about issuing purchase orders; it is an* ***end-to-end process*** *that links contract review, technical submittals, vendor management, logistics, and site readiness. Any weakness in this chain leads to* ***delays, rework, or cost escalation.***

5.9.6 Deliverables of Procurement Planning

- Procurement Schedule (integrated with project schedule).
- Approved vendor list.
- Procurement tracker.
- Risk register updates.

5.9.7 Procurement Risks vs Mitigation in HVAC Projects

Table 5.16 Procurement Risks vs Mitigation in HVAC Projects

Risk	Impact on Project	Mitigation Strategy
Delay in approval of Drawings	Delays in drawing approval cause delays in arriving at the take-off quantities and arriving at HVAC equipment specifications	Submit the drawings early for approvals. Finalise the Plant room location, AHU locations, pipe routings, and ducting routings.
Delay in approval of technical submittals	Delays in placing purchase orders → long-lead items (chillers, AHUs, pumps) not available on time.	Submit Technical submittals early, track status with a Submittal Tracker, and follow up regularly with the consultant/client.
Vendor lead times are longer than expected	Project milestones slip, and downstream activities get delayed.	Identify long-lead items during planning; place POs early; negotiate phased deliveries.
Import clearance delays (customs, port handling)	Equipment held at port → site progress stalled.	Appoint a logistics consultant; prepare documents in advance; allow a buffer in the schedule for imports.
Limited vendor options / single-source supply	Vendor has pricing power; delays cannot be easily substituted.	Pre-qualify multiple vendors during initiation; keep at least one backup supplier approved.
Poor quality or non-compliant material supplied	Rejections, rework, disputes, and project delays.	Insist on factory inspections (FAT), QA/QC checks on delivery; keep compliance with specs non-negotiable.
Bulk ordering without storage readiness	Rusting, theft, site congestion, and rework.	Align deliveries with site readiness; secure storage space; stagger deliveries as per installation sequence.

Risk	Impact on Project	Mitigation Strategy
Material theft/loss at the site	Cost overruns, repeat procurement.	Secure storage area with access control; maintain material registers; deploy security for high-value items (e.g., copper pipes).
Payment disputes with vendors	Vendors hold dispatch → delivery delayed.	Align clear commercial terms upfront; release payments as per agreed milestones; maintain transparency.
Outdated drawings used for procurement/ fabrication	Rework, wastage of material, and added cost.	Always procure/fabricate against the latest approved drawings only; maintain a drawing tracker.
Coordination gaps between procurement & site teams	Wrong items delivered or delivered too early/late.	Weekly coordination meetings; integrate procurement schedule with construction master plan.

5.9.8 Procurement Tracker

There are many tools available for procurement tracking purposes. ERP systems generally are more reliable as they have built-in track POs processes. They offer dashboards, too. A typical procurement tracker is given below. Please note the following for effective project tracking

- Procurement Schedule linked to the master schedule.
- Tracker columns: Item, Qty, Vendor, Submittal Status, PO Date, Delivery Date, Actual Receipt, Remarks. You may also mention FAT requirements, delivery challans
- Responsibility: Procurement Team + PM.

Golden Rule – Procurement - "Every delay in procurement today becomes a delay in installation tomorrow."

5.9.9 Work Orders for Subcontracted Works

In addition to procuring equipment and materials, HVAC projects depend heavily on **subcontracted services**. These include:

- Equipment installation (chillers, AHUs, VRFs, pumps, cooling towers)
- Duct fabrication and installation
- Pipe laying and insulation
- Electrical panel installation and cabling works
- Testing, adjusting and balancing works
    - Miscellaneous works (shifting of materials, civil foundations, wall openings, equipment supports, scaffolding, painting)
    - Site preparation, security works

Importance of Work Orders

Work Orders (WOs) formalise the scope, cost, and schedule for subcontracted works. Unlike informal job assignments, a WO ensures accountability, quality, and timely execution.

A good Work Order should include:

1. Scope of Work – Clearly define what is included (and excluded).

2. Reference Documents – Drawings, BOQ, specifications, and method statements.

3. Quality Requirements – Standards to be followed (e.g., SMACNA for ducting, IS codes).

4. Safety Requirements – Site safety, PPE, permits, work-at-height requirements.

5. Schedule and Milestones – Start date, completion date, and intermediate deadlines.

6. Payment Terms – Linked to measurable deliverables.

7. Responsibilities– Contractor vs client vs subcontractor roles.

8. Penalties/Delays– Clauses for delays, rework, or non-compliance.

Benefits of Work Orders in HVAC Projects

- Avoids scope disputes with subcontractors.
- Provides cost control and visibility.
- Enables progress monitoring against milestones.
- Ensures subcontractor alignment with the overall project schedule.
- Reduces delays caused by ambiguity or miscommunication.

Golden Rule – Subcontracting Work Orders: *"A Work Order is not paperwork — it is the contract on which your site's progress depends. Release the work order before starting the work."*

Table 5.17 Dos and Don't of Procurement Planning

✅ Dos	❌ Don'ts
• Prepare a **procurement plan early**, linked to schedule milestones (long-lead items like chillers, AHUs, VRFs).	• Don't wait until site execution to place POs — long-lead delays cripple the project.
• Cross-check BOQ vs. approved shop drawings before releasing purchase orders.	• Don't release POs based only on tender BOQ without updated quantities.
• Use approved vendor lists and ensure all materials match contract specifications.	• Don't source from unapproved vendors unless formally approved.
• Get technical submittals approved before ordering.	• Don't bypass consultant/client approval for urgent deliveries.
• Include **clear delivery terms, warranties, and payment clauses** in POs.	• Don't ignore Incoterms, insurance, or customs clearance requirements.
• Maintain a procurement tracker with status (submittal, PO, FAT, shipment, delivery).	• Don't depend on email trails only — use structured logs.
• Inspect materials at dispatch (FAT/inspection) and upon receipt at the site.	• Don't assume factory compliance = site compliance.
• Keep a buffer for import/customs clearance.	• Don't assume optimistic timelines — customs delays are common.
• Align procurement with the **cash flow plan**.	• Don't overstock the site with excess material — risk of wastage and disputes.

5.9.10 Summary – Procurement Planning in HVAC Projects

Procurement is the lifeline of HVAC project management, with nearly 60% of project costs tied to materials. The success of a project depends not only on ordering the right equipment but on timing, coordination, and quality assurance throughout the supply chain.

In this chapter, we examined:

- The scope of HVAC procurement (**high side vs low side equipment**).
- Why HVAC procurement is inherently complex (engineered-to-order equipment, long lead times, vendor limitations).
- The procurement process flow, from contract review to material receipt.
- Risks in procurement (delays, theft, quality issues, logistics challenges) and their mitigations.
- The importance of tools such as procurement trackers, submittal logs, and risk registers to ensure visibility and control.
- Practical insights, case studies, and Golden Rules that reflect the realities of the Indian HVAC industry.

Closing Thought

Procurement teams are often invisible when things run smoothly, but when delays occur, the entire project feels the shockwave. Effective procurement is not just about buying — it is about **planning, negotiating, coordinating, and delivering certainty** to the project.

If project initiation sets the direction, and WBS defines the scope, then procurement ensures the physical backbone of the HVAC system arrives on time and within budget. A well-managed procurement process is therefore the difference between a project that struggles and one that succeeds.

Procurement is where project schedules are won or lost.

Checklist for Procurement Planning

- ✓ [ ] Review contract, BOQ, and approved drawings.
- ✓ [ ] Identify long-lead items (chillers, VRF, AHUs, pumps, panels).
- ✓ [ ] Align procurement schedule with project milestones.
- ✓ [ ] Prepare and submit **complete technical submittals** (no gaps).
- ✓ [ ] Shortlist vendors (technical, commercial, supply capability).
- ✓ [ ] Prepare **vendor comparison sheet** (tech + delivery + commercials).
- ✓ [ ] Negotiate and issue **Purchase Orders (POs**).
- ✓ [ ] Release Work Orders for Contractors.
- ✓ [ ] Monitor vendor progress, FAT, and inspections.
- ✓ [ ] Plan and coordinate **logistics & customs clearance.**
- ✓ [ ] Receive, inspect, and store materials securely.

Tip for Readers: Acknowledge your procurement team not just when there are problems, but also when deliveries are on time. Their silent success is what keeps your project on track.

5.10 Project Quality Management

"Quality is never an accident; it is always the result of intelligent effort." – John Ruskin

5.10.1 Introduction

Quality is one of the least understood aspects of project execution. Many team members wrongly associate it with ***endless documentation***, leading to the impression that quality is nothing more than paperwork. This misunderstanding causes the true purpose of quality to be missed entirely.

Quality is **not paperwork**. It is **not just an inspection**. It is **not an activity to be carried out after work is completed**. Nor is it merely corrective action once defects are found. **Quality is the responsibility of every member of the project team**, not just the QA/QC engineer.

Quality is what transforms design intent into delivery reality.

As the saying goes:
"Quality is realised only by its absence in a work product."

When systems fail, leak, or underperform, everyone notices quality — but by then, it is too late. The true goal of quality planning is to make quality visible and consistent from the very beginning.

Typically, organisations have a well-defined **Quality Manual,** with every quality activity well defined and documented. It documents the "quality assurance activities and Quality control activities separately for each process.

The Quality manual details:

a. Relevant standards, process flows, templates, and checklists
b. Safety procedures, IEQ parameters, life cycle & O&M procedures
c. Testing, validation, and acceptance procedures

5.10.2 Key distinction between Quality Assurance and Quality Control

- **Quality Assurance (QA):** Focuses on the **processes** to be followed to achieve the overall quality of the project.

- **Quality Control (QC):** Focuses on the **product,** verifying whether the delivered work or material meets the specified requirements.

Key differences between Quality Assurance and Quality control activities are enumerated in the following table 17 QA vs QC below

Table 5.18 QA vs QC

Aspect	Quality Assurance (QA)	Quality Control (QC)
Definition	Proactive process to prevent defects.	Reactive process to identify defects.
Objective	To assure quality by building it into the process.	To control the quality of products/ services by detecting deviations and enforcing corrections.
Focus	Processes – ensuring correct methods, planning, and documentation are in place. Necessary training and tools are available.	Outputs/Workmanship – inspecting completed work and materials.
Timing	Done before and during execution.	Done after or during execution.
Responsibility	Entire project team (PM, engineers, supervisors, vendors).	QA/QC engineer, site inspectors, consultant/client representatives.
Examples in HVAC	Preparing method statements. Technical submittals, scheduling inspections, and training supervisors. Regular Audit to check whether the process is followed	Duct leakage tests, hydrostatic pressure tests, insulation thickness checks, and chiller performance tests.

5.10.3 Quality Assurance Process Flow

The following **Table** 18 gives the Quality Assurance process flow:

Table 5.19 Quality Assurance Process Flow

Inputs:	**Process:**	**Outputs:**
1. Organisation policy document 2. Contract document 3. Various standards	1. Prepare QA plan for every process, including procurement, installation, acceptance and rejection criteria 2. Quality Audits: planning and frequency of audits 3. Corrective Action and Preventive Action plan Identify commissioning activities	1. Detailed Quality Assurance and Audits plan 2. Integrated Quality schedule with the Project schedule

5.10.4 Quality Control Process Flow

Quality Control plans are prepared for all the major equipment used under the project. Ducting, piping, AHU Installation, Chillers, DX systems, VRF systems, Pumps, cooling towers, FCUs, BMS and testing of all equipment

Quality Control Process flow is detailed in the Table 19 below:

Table 5.2018 Quality Control Process Flow

Inputs:	Process:	Outputs:
Approved drawings & technical specifications • BOQ (Bill of Quantities) •Material submittals & OEM datasheets • QA Plan & ITPs • Organisation's quality policy & standards (ISHRAE, ASHRAE, SMACNA, IS codes)	•Conduct inspections & tests (duct leakage, hydrostatic pressure, insulation checks, equipment performance) • Verify workmanship against approved method statements • Use calibrated instruments (flow meters, micrometres, gauges) • Raise & track Non-Conformance Reports (NCRs) • Conduct corrective & preventive action follow-ups (CAPA) •Record inspection/test results in standard formats.	Inspection & Test Records (ITRs) • NCR logs & closure reports • Performance test certificates • As-built drawings with QC approvals • Compliance to acceptance criteria • Verified & validated HVAC system ready for commissioning

Golden Rule – Quality Planning & Performance: *"Quality is not proven when equipment runs — it is proven when the whole system performs as designed, in every season."*

5.10.5 Quality Planning

Quality planning should deliver measurable outcomes:

- HVAC system performance criteria
  - Temperature, Humidity, Air filtration level, Efficiency of the subsystems
- Indoor Air Quality and IEQ (Indoor Environment Quality)
- Water usage and potential reduction

- Greenhouse Gas emissions reduction
- Sustainability

Case Study – Good Quality Planning
Project: Corporate IT Park – Bengaluru

Scenario: The HVAC contractor implemented a detailed Quality Assurance Plan from day one. All technical submittals were cross verified, a duct leakage test was performed for every zone, and method statements were followed rigorously. The project team conducted weekly QA/QC meetings with the consultant, and mock-ups were created for duct insulation and AHU installation before full-scale work began.

Outcome:

- Minimal rework during execution.
- Timely approvals of inspections
- Final commissioning achieved 3 weeks ahead of schedule.
- Client satisfaction was high, resulting in early release of payments and an extended service contract.

Lesson Learned: Proactive quality planning avoids rework and builds client trust.

Deliverables of Quality Planning

- Quality Assurance Plan (QAP).
- Inspection & Test Plan (ITP).
- Quality Checklists.
- Approved method statements.
- Quality logs and registers.

5.10.5.1 Quality Planning Process

Quality planning is carried out in the following steps:

Step 1. Review contract specifications: The first step is to review contract specifications, BOQ and other special conditions for all the major equipment. This may be compared with the organisation's Quality policy. Based on the requirements, develop a Quality Assurance Plan (QAP) for all the processes, materials and systems. Number of audits required in each stage of the project, and resource requirements for the audits.

Step 2. Prepare Inspection & Test Plans (ITPs): This is an important step. Inputs for this process are the basis of the design document, DBR, Technical specifications, BOQ, approved drawings, approved technical submittals and method statements.

Inspection plans are to be prepared for all the materials/ equipment that are used in the project. A sample board is normally prepared and kept at the site for quick reference. Define **acceptance criteria** (e.g., duct leakage, pump efficiency, insulation thickness).

Inspection plans are prepared for:

- Pumps
- Cooling Tower
- Air distribution products
    - Ducting
    - VAV
    - Grills, diffusers, dampers

  - Fire dampers
  - Canvass connections
  - Other accessories
- **Piping and accessories**
  - Chilled water pipes
  - Condenser water pipes
  - Chemical dozing
  - Valves (Butterfly valves, Balancing valves, PIBC valves, NRVs, Ball Valves and other valves as mentioned in the BOQ)
- **Insulation**
  - Duct Thermal Insulation
  - Acoustic Insulation
  - Chilled water piping insulation
  - Plant room/AHU room insulation
- **Air Separator, Expansion Tank**
- Electrical panel
- Cables, Cable Trays
- BMS System and other system components based on the specifications.

Quality requirements for components (ducts, pipes, insulation, pumps, chillers).

- Test Certificates, Inspection reports, Test results.
- Workmanship quality standards.
- Documentation requirements (inspection reports, test certificates, calibration certificates).

Step 3 – Plan For Health Safety, and Environment

A well-thought-out Health, Safety, and Environment (HSE) plan is crucial for ensuring the successful completion of an HVAC project without any harm to personnel, property, or the surrounding environment. The plan's objective is to systematically identify, evaluate, and control hazards associated with all phases of HVAC installation, commissioning, and maintenance.

2. Scope

This HSE plan applies to all HVAC activities at the project site, including but not limited to: installation, equipment handling, ductwork, pipework, insulation, welding, use of cranes, and work at height. All personnel, including contractors and visitors, must comply with the protocols outlined. The Project Manager and the site team should lead the way.

3. Responsibilities

- **Project Manager**: Ultimate responsibility for HSE compliance, resource allocation, and reporting incidents lies with the Project Manager. There is always a temptation to bypass the safety protocols for getting things done quickly. This should be avoided by all means.

- **HSE Officer**: Identify one well-trained and qualified safety officer for on-site monitoring, regular inspections, risk assessments, training, and incident investigations.

- **Site Supervisors**: Site supervisors should be trained and authorised to enforce safe work practices and intervene in unsafe activities.

- **Workers**: All contract workers must comply with safety instructions and use necessary PPE and tools.

4. General HSE Measures

- ✓ Display safety posters and HSE instructions at prominent locations.
- ✓ Conduct a Job Safety Analysis (JSA) before each task.
- ✓ Ensure Safety Data Sheets (SDS) for all chemicals are on site and personnel are trained to find and understand them.
- ✓ Emergency numbers, evacuation plans, and first aid information must be displayed prominently at different locations at the site.
- ✓ Maintain records of all HSE briefings, inductions, and toolbox talks.

5. Safety Tools, Equipment, and Personal Protective Equipment (PPE)

- **Essential PPE**: Plan for necessary Hard hats, safety goggles, gloves, high-visibility vests, steel-toed boots, ear protection, flame-resistant clothing (for hot work/welding), and respiratory protection when handling chemicals (e.g., refrigerants). These should be available in the Stores at the site, under the supervision of the HSE officer.

- **Safety Tools**: Plan for sufficient numbers of Digital multi-meters, gas leak detectors, airflow meters, thermometers, and heavy-duty drills as standard tools in addition to PPE. Ensure these are in working condition and calibrated regularly.
- **Specialised Equipment**: Wherever required, ensure lifting slings, shackles, chain blocks, cranes rated for load, scaffolds, ladders, and fall arrest systems (harness and lanyard) are available.

6. Posters, Signage, and Safety Communication

- ✓ Preparation of HIRA – Hazard Identification and Risk Assessment
- ✓ Display hazard signs at work areas (e.g., "Caution: Working at Height," "Welding in Progress," "Crane Operation").
- ✓ Emergency exit signs, muster point markers, and fire extinguisher locations clearly marked.
- ✓ Place instructions for the correct use of PPE and first aid procedures near work entrances.

7. Training and Competency

- ✓ Conduct site induction for every worker before beginning activities.
- ✓ Organise regular safety drills (fire, evacuation, spill, electric shock).

- ✓ Specific training for:
- ✓ Working at heights
- ✓ Crane operation and lifting
- ✓ Hot work (welding, cutting, bitumen handling)
- ✓ Confined space entry.eeslindia+2
- ✓ Maintain training records and ensure only trained/ authorised personnel conduct high-risk activities.

8. Working at Heights

- ✓ Generally, all work above 2 meters requires prior risk assessment.
- ✓ Use fall protection (full body harness with double hooks, safety nets, proper anchorage).
- ✓ Only trained personnel permitted to use scaffolds, mobile platforms, or ladders.
- ✓ All openings and edges must be protected with guardrails or covers.
- ✓ Use a permit-to-work system for any non-routine height work.

9. Usage of Cranes and Heavy Lifting

- ✓ Conduct a **risk assessment** and lift plan for each operation.
- ✓ **Inspect cranes for certification**, load rating, and physical condition before use.

- ✓ Only certified operators may run cranes; riggers/ signallers must be designated.
- ✓ **Exclusion zones** and **warning signage** established around lifting areas.
- ✓ **Never exceed rated load**; ensure proper rigging of loads.
- ✓ Weather conditions (wind, rain) assessed before and during lifting.

10. Hot Bitumen Insulation

- ✓ Use PPE, including flame-resistant clothing, face shields, and gloves.
- ✓ Work areas roped off with warning signage ("Hot Surface").
- ✓ Fire extinguishers and a burn first-aid kit must be on site.
- ✓ Handle bitumen only in well-ventilated areas.
- ✓ Store bitumen away from flammables and ignition sources.

11. Welding and Hot Work

- ✓ Hot work permits are required before welding, cutting, or grinding.
- ✓ Clear the area of combustibles or cover with flame-retardant blankets.
- ✓ Use **fire-resistant PPE**, including face and eye protection.

- ✓ **Designate a trained fire watcher during and after work.**
- ✓ **Equip work area with fire extinguishers**, sand buckets, and emergency cut-off for gas/electric supplies.
- ✓ Ensure proper ventilation, use of extraction fans where needed.

12. Environmental Safeguards:

- ✓ Segregate waste: metals, plastics, refrigerants, hot bitumen, and chemical containers must be disposed of according to regulations.
- ✓ Use drip trays/pans for oil and chemical handling.
- ✓ Minimise dust and noise by using barriers, wet methods, and scheduling noisy activities appropriately.
- ✓ Avoid work in extreme weather unless hazards are controlled (e.g., wind, rain, high temperatures)

13. Incident and Emergency Response

- ✓ Create an Escalation Matrix
- ✓ Prepare an Emergency Handling guide which includes – Guidelines, Hospital details, ambulance call numbers, mock drills
- ✓ Immediate first aid given; incident reported to HSE officer and recorded.
- ✓ Initiate evacuation and call emergency services for serious incidents.

- ✓ Investigate all accidents and near misses; implement corrective actions.
- ✓ Review and update the HSE plan continuously based on lessons learned and as project phases change

14. Monitoring and Review

- ✓ Conduct daily site inspections.
- ✓ Hold weekly HSE meetings and toolbox talks.
- ✓ Track and display safety performance metrics (e.g., days without accidents).
- ✓ Periodic HSE audits by third-party or corporate safety teams. scribd+1

This HSE plan ensures that HVAC projects are executed safely and with minimal environmental impact by integrating robust safety practices, continuous training, and strong supervisory oversight.

Note: These are only for informational purposes. Detailed and complete HSE plans to be prepared by Qualified HSE professional.

Table 5.2119 Dos & Don'ts – HSE in HVAC Project Execution

☑ Dos	❌ Don'ts
• Conduct toolbox talks and safety briefings before starting every shift. • Ensure all workers use PPE (helmets, safety shoes, gloves, harnesses, goggles). • Verify permits before hot works, working at height, or confined space entry. • Keep fire extinguishers, first-aid kits, and emergency exits accessible at all times. • Regularly inspect scaffolding, lifting equipment, and power tools for compliance. • Maintain site cleanliness — remove debris, sharp objects, and unused materials daily. • Report and document all near-misses, incidents, and unsafe practices immediately. • Provide safety training for new workers and refresher sessions for existing staff.	• Don't allow untrained or unauthorized personnel to operate machinery or tools. • Don't bypass lock-out/tag-out (LOTO) procedures during electrical work. • Don't overload scaffolds, ladders, or lifting equipment. • Don't allow welding, brazing, or grinding without protective barriers. • Don't block fire exits, emergency routes, or access to safety equipment. • Don't ignore consultant/client-specific HSE protocols — they are often stricter. • Don't rush tasks by cutting corners on safety to save time.

Golden Rule: *"No deadline is worth a life — plan safety as carefully as you plan cost and schedule."*

Step 4 – Identify Responsibilities (PM, QA/QC Engineer, Consultant)

While quality is everyone's responsibility, it is essential to **define clear roles and responsibilities for** quality assurance and quality control during the planning **stage** itself. Without this clarity, quality activities risk becoming informal and inconsistent.

- **Project Manager (PM):** Overall accountability for ensuring quality processes are embedded in project execution.
- **QA/QC Engineer:** Day-to-day responsibility for inspections, testing, documentation, and reporting deviations.
- **Consultant/Client Representative:** Independent verification and approval of submittals, inspections, and test results.
- **Supervisors & Technicians:** Execution of tasks as per approved method statements and checklists.
- **To make this effective:**
  - ✓ Team members must be **trained** to perform correctly.
  - ✓ They should be provided with the "**necessary tools, gauges, checklists, and techniques**" to verify quality.
  - ✓ Most importantly, they must have the authority to accept or reject materials, subsystems, or installations that do not meet specifications.
  - ✓ A clear and documented approval process for materials and installation methods should be maintained at the site, ensuring accountability and transparency at every step.

Golden Rule: *The quality team can only identify the defects. Defect rectification and prevention are the roles of the respective project/ site engineers.*

Step 5 – Plan for Third-Party Inspections and Commissioning

In many projects, it is common practice to invite a third-party commissioning team to participate in the commissioning process. Their role is to ensure independent, unbiased validation of system performance. ***This is not a mandatory process for every project.***

When engaging a third party, the following criteria should be considered:

- **Competence:** Must be technically qualified to complete the commissioning activities.
- **Relevant Experience**: Should have prior experience in commissioning equipment of a similar type and scale.
- **Proven Value Addition:** Must be able to demonstrate how their involvement has added measurable value in previous projects.
- **Resources and Tools**: Should be equipped with manpower, instruments, gauges, and calibration facilities required for the task.
- **Communication Skills:** Must be able to clearly document findings and communicate effectively with the project team.
- **Training:** Should be trained and certified for each commissioning activity and process.
- **Documentation:** Must have standard checklists, acceptance criteria, and reporting formats in place.

The scope of the third-party team should be clearly defined in advance — including both inclusions and exclusions — to avoid overlap or ambiguity.

Note: The ultimate responsibility for quality always rests with the Project Team. A third party's role is to enable and validate quality within the boundaries of their defined scope, not to replace the project team's accountability.

> **Golden Rule:** *"Third Party Commissioning "Third-party commissioning adds value, but responsibility for quality always remains with the project team."*

Step 6–Monitoring and Control of Quality Activities

Monitoring and control are one of the most critical aspects of quality planning. Many projects fail not because of poor design or procurement, but due to the lack of rigour in tracking and controlling quality on site.

A common pitfall is when the QC Team either micromanages activities or conducts reviews that are merely procedural, without focusing on the quality intent.

> **Golden Rule:** The QC team must understand the purpose, intent, and process behind every activity to make quality management truly effective.

5.10.6 HVAC System Commissioning

"Commissioning is more than just switching on the HVAC equipment."

Commissioning validates that the facility meets the OPR. It confirms that components and subsystems perform in harmony and must be integrated with QAP and the schedule. Commissioning is not just switching on equipment; it is system-level verification and seasonal validation. (Dedicated commissioning chapter follows later in the book.)

It ensures that:

- Every component performs as per specification.
- All subsystems work in harmony.
- The system delivers the expected comfort, energy efficiency, safety, and reliability. It is achieved not just on day one, but throughout its lifecycle.
- To achieve this level of system effectiveness, a strict and consistent quality control mechanism is essential at every stage — not only at the end.

Quality control must be integrated into the HVAC Commissioning process from the very beginning — starting in the design and planning phase, continuing through procurement, installation, testing, and finally commissioning.

“Commissioning is not the end of quality — it is the proof that quality was present at every step, testing the integration of these components within the system.

Note: A dedicated chapter on HVAC commissioning is provided later in this book.

5.10.7 System-Level Quality in HVAC Projects

The quality of an HVAC system can be impacted by several factors:

- Incorrect or incomplete design.
- Ambient weather variations depending on the time of day.

- Incorrect load assumptions (lighting, occupancy, equipment load).
- Errors in procurement or non-compliant materials.
- Poor installation practices.
- Changes at the site after design has been finalised (e.g., adding extra AHUs, extending air-conditioning to new areas).

HVAC systems are generally designed for peak ***loads***, with reasonable ***safety margins*** built in at each stage. These peak conditions occur only **1–3%** of the time in a year.
This design creates a false sense of comfort during testing and handover.

- At regular load conditions (Normally less than the Peak loads and exist most of the year), the system easily delivers temperatures that appear even better than design.
- This makes it deceptively easy for contractors to hand over the system without addressing underlying inefficiencies.
- Problems such as poor balancing, incorrect load assumptions, or integration issues often remain hidden until the system is in continuous operation.

The lesson: Overdesign may mask quality issues at handover, but these inefficiencies always surface later in the form of high **energy bills, poor comfort, or recurring complaints**.
"An HVAC system that looks good on day one may only be hiding flaws — overdesign makes handover easy, but operation difficult."

In essence, quality in HVAC projects is a continuous process, focused on how well all subsystems — air distribution, hydronics, controls, and equipment — work together to achieve optimum system performance.

5.10.8 Key Elements of Sub-System-Level Quality

In HVAC projects, the performance of individual subsystems at their **design parameters** — and the correct **balancing of these subsystems** — is critical to achieving overall system efficiency.

For example, a primary chilled water pump may be designed to operate at a specific head and flow rate, with an expected efficiency of 75%. The motor is sized to match this requirement. If the head is increased or the flow rate is altered, the pump efficiency changes, and the power requirement increases, leading to energy wastage and operational inefficiency.

1. **Airflow Testing and Balancing (TAB)**

   - Measure and verify airflow at AHUs, FCUs, diffusers, and grilles.
   - Ensure distribution matches design volumes.
   - Balance system to eliminate hot/cold spots.

2. **Hydronic Testing and Balancing**

   - Verify flow rates in chilled water and condenser water circuits.
   - Check pump head and efficiency against design.
   - Balance valves and flow meters for correct distribution.

3. Chiller and Equipment Performance Testing

- Confirm chiller COP/EER matches design at full and part load.
- Monitor VRF performance across multiple indoor units.
- Evaluate cooling tower performance (approach temperature).

4. System Integration with Controls (iBMS)

- Verify sensors, actuators, and control logic.
- Check alarms, set points, and energy meters.
- Confirm sequences of operation (start-up, shutdown, load variation).

5. Seasonal and Part-Load Performance

- Evaluate the system under varying conditions: peak summer, monsoon humidity, and winter loads.
- Ensure energy efficiency targets are met across operating ranges.

6. Documentation and Handover

- Maintain commissioning records, test certificates, and calibration logs.
- Prepare O&M manuals and train facility staff.

> ***Golden Rule – System Quality:*** *"A HVAC system is only as good as its performance under real conditions, not just at installation."*

5.10.9 Quality Tools in HVAC Projects

To make quality planning effective on-site, use a small set of dependable tools that enforce consistency and traceability. Some of the tools used are listed below.

- **Checklists** — Simple, task-based (Ducting, Piping, Insulation, Chiller positioning). Carry them on site for inspections and sign-offs.

- **Inspection & Test Plans (ITPs)** define what, when, who, acceptance criteria, and record. Use ITPs for every major deliverable (ducts, piping, chillers, AHUs).

- **Mock-ups & Samples** — trial installations (e.g., anchor bolts & chiller base, duct insulation mock-up) to validate workmanship before full rollout.

- **Calibration Logs** — maintain calibration records for instruments (manometers, anemometers, temperature probes, flow meters). Include certificate expiry dates and next-calibration reminders.

- **Non-Conformance Reports (NCRs)** standardised form for recording deviations, corrective & preventive actions, ownership and closure date.

- **Material Acceptance Records** — signed delivery and inspection records (GRNs with inspection fields) for GI, copper, insulation, pumps, electrical panels, etc.

- **Quality Review Reports** — periodic summary (weekly/monthly) of inspections, NCRs, recurring issues, corrective actions, and lessons learned.

- **Mock Commissioning Checklists** — pre-commissioning checks (valve positions, sensors, test points, flushing completed) to avoid last-minute firefighting.

- **Training & Competency Records** — site personnel training logs (e.g., brazing, rigging, TAB methods) and certification copies for quick verification.

- **Visual Evidence** — standard photo log naming convention (date_area_equipment) to store key milestone photos with captions (helps in claims and dispute resolution).

5.10.10 Risks of Poor-Quality Planning

- Rework and delays.

- System inefficiency (high energy costs, comfort issues).

- Client dissatisfaction and payment disputes.

- Long-term O&M problems.

Golden Rule – Quality Planning - *"Rework costs more than doing it right the first time — plan quality, don't patch it later."*

5.10.11 Best Practices in Quality Monitoring & Control

1. Conduct Detailed Quality Reviews:

- Identify **Corrective Actions (CAs) and Preventive Actions (PAs).**

- Capture and share best practices across teams.

2. Integrate Quality in Project Reviews:

- Present a **Quality Summary** as part of regular project meetings.

- Ensure stakeholders act on identified risks, deviations, and issues related to cost and timelines.

3. Protect Quality from Project Pressures:

- It is the Project Manager's responsibility to ensure quality is not compromised due to time or cost pressures.
- Substandard materials, workmanship, or shortcuts should ***never be accepted***.

4. Action Tracking:

- Circulate minutes of meetings (MoM) with clear action items.
- The quality team must follow up on the closure of these actions within agreed timelines.
- Any risks or schedule/budget impacts arising from quality actions should be flagged early and managed appropriately.

Golden Rule – Monitoring Quality: *"Filling checklists is not monitoring; acting on risks, deviations, and lessons learned is what makes quality real."*

5.10.12 ☑ Dos and ✖ Don'ts of Quality Planning

Table 5.2220 Dos & Don'ts – HSE in HVAC Project Execution

☑ Dos	✖ Don'ts
Prepare a **Quality Assurance Plan (QAP)** and **Inspection & Test Plans (ITPs)** for each system (ducting, piping, insulation, equipment).	Don't start installation without approved QAP/ITPs.
Follow **approved method statements** and standards (SMACNA, ISHRAE, ASHRAE).	Don't accept "site practices" as substitutes for standards.
Conduct **mock-ups/sample approvals** for ducts, piping, and insulation.	Don't skip mock-ups — they prevent large-scale rework.
Maintain **calibrated instruments** for inspections (micrometres, gauges, flow meters)	Don't use uncalibrated tools — results will be disputed.
Document every inspection with **signed reports & checklists**.	Don't rely on verbal clearances — "if it isn't documented, it doesn't exist."
Engage **third-party testing** for critical systems when required.	Don't assume internal checks alone will satisfy consultants/clients.
Train supervisors and technicians on quality procedures.	Don't assume QA/ QC engineer alone is responsible for quality.
Ensure **non-conformance reports (NCRs)** are tracked with corrective/preventive actions.	Don't close NCRs without root cause analysis.
Link **quality milestones** with payment terms and commissioning.	Don't treat quality as paperwork at the end of the job.

5.10.13 Conclusion – Quality Planning in HVAC Projects

Quality in HVAC projects is often misunderstood as paperwork or an end-of-project inspection. Quality **is the invisible thread that runs through every stage of the project** — from design intent and procurement to installation, commissioning, and handover.

Delivering quality consistently builds a reputation faster than winning a single big job. A contractor who **plans quality, documents it, trains people, and proves it during commissioning** becomes predictable, reduces claims, retains clients, and wins repeat business. In the Indian HVAC market, reputation is earned by doing the small things right every time the mock-up, the calibrated gauge, the signed material inspection, and the timely NCR closeout. That is professional excellence and the safest route to long-term growth.

Annexures

Annexure 1. Acceptance Criteria for Materials in HVAC Projects

For effective quality control, the *acceptance criteria* of each material should be clearly specified at the procurement and inspection stage. Relevant Indian ***Standards (IS), International Standards (SMACNA, ASHRAE), and project-specific BOQ/technical specifications*** should be referred to define these criteria. Necessary tools to inspect and measure should be calibrated and made available at the site/ stores.

Example: Ducting Material
Base Standard: Galvanised steel sheet as per **IS 277 (zinc coating grade).**

Sheet Thickness: Varies by application — determined by IS 655 (for ducts) or SMACNA guidelines.

Fabrication & Installation: Jointing, supports, reinforcements, and sealing must conform to IS 655/ SMACNA.

Material Acceptance Checklist (Illustrative) is shown in Table 4 below.

Table 5.23 -Illustrative Material Acceptance Checklist

Material	Acceptance Criteria	Reference Standard	Inspection Method	Record
GI Sheets for ducts	Thickness as per IS 655 / BOQ; Zinc coating (ZSM) as per IS 277	IS 277, IS 655, SMACNA	Micro-meter, Coating Gauge	Material Inspection Report
Copper Tubes	Diameter, wall thickness, temper; free of dents/cracks	ASTM B88 / IS 10773	Vernier calliper, Visual	Material Log
Pumps	Flow rate, head, efficiency; nameplate matches BOQ	IS 9137 / OEM datasheet	Factory Test Certificate	QA Record
Insulation	Thickness, density, thermal conductivity	IT IS 3346, ASHRAE	Thick-ness gauge, Weighing	Inspection Log
Electrical Panels	IP rating, busbar size, breaker ratings	IS 8623	Visual, Test Certificate	Inspection Report

Recommendation:

A Material **Acceptance Checklist** should be prepared and used at the site. Every delivery must be **inspected and recorded** before materials are taken into inventory. This ensures only compliant materials enter the project, reducing rework and disputes downstream.

Golden Rule – Material Acceptance: *"If it is not inspected, it is not accepted."*

Quality At a Glance:

Table 5.24 - Quality at a Glance

Topic	Key Points / Action
Quality Intent	Deliver HVAC systems that meet OPR: comfort, IEQ, energy efficiency, and reliability.
QA vs QC	QA = processes (plan, train, audit). QC = outputs (inspections, tests, NCRs).
Top Deliverables	QAP, ITPs, Method Statements, Material Acceptance Records, Calibration Logs, NCR register, Commissioning records, O&M manuals.
Must-do Steps	1. Review contract & standards. 2. Prepare QAP & ITPs. 3. Appoint QA/QC owner. 4. Calibrate instruments. 5. Run mock-ups. 6. Do pre-commissioning checks. 7. Conduct seasonal performance tests.
Key Tools	Checklists, ITPs, Mock-ups, NCRs, Calibration logs, Material Inspection Reports, Photo logs, and QA Review reports.
Part-load Warning	Systems often hide problems at handover. Test & tune for part-load and seasonal conditions.
Third-party checks	Use independent commissioning/testing for high-risk items (FATs, water quality, coating tests).
Golden Rules	"Inspecting quality at the end is too late." "If it is not inspected, it is not accepted."
Quick Measures	Weekly QA summary, immediate NCR closeouts, keep instrument calibration certificates on site, and train supervisors in ITP usage.

5.11 Project Risk Management

"Risk comes from not knowing what you are doing." – Warren Buffett

5.11.1 Introduction to Risk Planning

What is Risk?

Risk is an uncertain event that may or may not occur, but if it does, it has the potential to affect project deliverables and **disrupt costs**, schedule, or quality. Importantly, not all risks are negative — some may **present opportunities** to improve outcomes.
When assessing a risk, the following factors must be considered:

- ✓ **Probability**: How likely is it that the risk will occur?
- ✓ **Impact:** What effect will it have on cost, schedule, or quality?
- ✓ **Timing**: When is it most likely to occur during the project life cycle?
- ✓ **Frequency:** How often could it occur?
- ✓ **Detectability:** How easy is it to identify when the event is happening?

5.11.2 Why Risk Planning is Crucial in HVAC Projects

Every HVAC project carries uncertainty. With multiple touchpoints — design, procurement, installation, commissioning, and coordination with several agencies — risks are inevitable. If left unplanned, they can delay schedules, escalate costs, and compromise quality.

5.11.3 Typical categories of risks in HVAC projects include:

- **Technical Risks** – design errors, incorrect equipment sizing, clashes in drawings.
- **Resource Risks** – labour shortages, lack of skilled welders, delays in subcontractors.
- **Environmental Risks** – floods, extreme weather, site access issues.
- **Safety Risks** – accidents during lifting, welding, or working at heights.
- **Commercial Risks** – disputes in BOQ quantities, currency fluctuations, and contractual delays.

5.11.3.1 Risks can be classified into two broad groups:

- **External Risks** – outside the organisation's control (regulatory changes, market dynamics, government policies, extreme weather). These may occur at any stage and require contingency reserves.
- **Internal Risks** – within the organisation's control, if identified early. These include risks related to scope, cost, schedule, human resources, materials, processes, technical issues, equipment, safety, and commercial disputes. With proper planning, many of these can be mitigated or avoided.

The following table, Table 26: Risk Impact table, shows tentative risks with their impact

Table 5.25 - Risk impact table

Risk category	**Risk Description**	**Impact**
Internal Risks		
Technical	Pump selection does not meet the design requirements	The pump may need replacement. It could impact the schedule and cost.
Commercial	1. Dispute in the Measurement of Pipes and ducts between the customer and the contractor. 2. Long lead items ordered without technical data sheet approval.	1. There are standard methods of Measurement. However, if they are not specified correctly, it could impact on COST. 2. High side items are manufactured to order. Any changes subsequently will disrupt the cost and delivery schedule.
Procurement	Insulation material was short-ordered.	Reordering and delivery to the site in small quantities may add to the cost and waste resources time at the site
Quality	An inexperienced worker did welding. Rejected by the customer/ PMC.	This will impact the schedule and rework cost due to wastage
Safety	Safety protocols were not followed. Safety is compromised.	It could lead to serious injuries to the personnel/ materials.

External Risks		
Regulatory	New taxes are proposed on some of the raw materials used	Impact the availability of those materials and the cost
Market	Copper prices internationally increased due to Dollar price fluctuations.	It could seriously impact the costs
Environmental	Flash floods and incessant rain inundated the AHU factory.	Delays in the schedule

5.11.4 Difference between issues and risks

Several people get confused between risks and Issues. A clear distinction is mentioned in the following table 2: Issues vs Risks

Table 5.26: Issues Vs Risks

Issues	Risks
The issue is a present reality at hand. This needs to be addressed immediately. It requires an experienced and responsible manager to address and resolve without being escalated. It has the potential to impact the project if not addressed at the earliest.	A risk may or may not occur. It is a futuristic uncertainty. Its impact may be severe and hence needs to be mitigated appropriately by making a risk response plan.
Issues are updated in the ISSUE LOG register and reviewed at regular intervals until resolved.	Risks are updated in the Risk register, plan for mitigation and then monitored closely to mitigate even before they occur or minimise the impact of occurred.

5.11.5 Risk Identification

The success of any HVAC project depends heavily on **identifying as many risks as possible in advance** and then tracking them closely throughout the project. Risks are often linked to specific functions, so the respective managers or teams are best placed to identify them.

A few examples are explained below to identify the risks.

Design Team: Can highlight risks such as delays in preparing technical submittals, incomplete information from the client, frequent design changes after drawings are finalised, or slow consultant approvals.

Procurement Team: May identify risks such as customs delays for imported items, damage during transit, or incomplete/incorrect assemblies being dispatched.

Site Inventory/Logistics Manager: Might raise risks such as a lack of secure storage space, improper handling facilities, or unavailability of inspection tools.

Project Manager/PMO: Should consolidate risks suggested by all stakeholders into a single Risk Register.

Tip: ***Never underestimate or discard a risk suggested by the team. What appears minor at the start may escalate into a major issue later.***

5.11.6 ☑Do's & ✖Don'ts of Risk Management in HVAC Projects

Table 5.27 Do's & Don'ts – Risk Management in HVAC Projects

☑ Dos	✖ Don'ts
• Keep a live Risk Register — review it weekly with the team. • Encourage all stakeholders (engineers, supervisors, vendors) to voice potential risks. • Link risks to actions — every risk must have an owner and response strategy. • Use lessons learned from past projects to pre-identify risks. • Maintain contingency reserves for known risks.	• Don't treat risk management as a paper exercise at project start. • Don't ignore minor risks — today's minor leak can become tomorrow's shutdown. • Don't delete resolved risks from the register — archive them for knowledge sharing. • Don't assume consultant/client will automatically cover unforeseen risks — check contracts. Don't wait until issues explode — escalate early.

5.11.7 Tools for Risk Identification

Several tools and techniques can be used to systematically identify risks

- **Brainstorming Workshops** – Cross-functional sessions with project stakeholders. Involve domain experts in discussions and identify potential risks for the projects

- **Checklists** – Based on past projects, lessons learned, and organisational knowledge assets.

- **Expert Judgement:** Expertise should be considered from individuals or groups with Specialised knowledge in the following topics:
    - Familiarity with the organisation's approach to managing risk
    - Tailoring risk management to the specific needs of a project
    - Types of risks that are likely to be encountered in projects in the same area
- **SWOT Analysis** – Identifying Strengths, Weaknesses, Opportunities, and Threats.
- **Data Analysis**
    - Stakeholder analysis to determine risk appetite of stakeholders.
    - Budget
    - Risks from previous projects, Historical Record and Reviewing documentation from similar projects to uncover recurring risks.

At the end of this phase, risk register is updated with all the identified risks. A summary report of all identified risks is circulated to all stakeholders.

Golden Rule – Risk Identification: *"A risk not captured well in advance is a problem waiting to happen."*

5.11.8 Risk Assessment and Prioritisation

Once risks are identified, the next step is to assess their significance and decide how they should be prioritised. Risk assessment is carried out using both qualitative and quantitative methods.

- **Qualitative Risk Assessment:**

Qualitative assessment is the first step, where risks are analysed based on their likelihood of occurring and their potential impact.

Key features of Qualitative Risk Management:

- Ensures risk data is unbiased and accurate.
- Evaluates probability and consequence of each risk using simple rating scales (e.g., High / Medium / Low).
- Provides a rapid and cost-effective way to prioritize risks.
- Forms the foundation for more detailed quantitative analysis.

Example in HVAC projects:

- Delay in chiller delivery or duct clashes with structural beams → **High Risk**.
- Site clearance delays → **Medium Risk**.

Outcomes: The outcome of qualitative assessment is an **updated Risk Register**, which includes:

- Description of the basis for each risk's probability and impact.

- Risks ranked by relative priority.
- Grouping risks by categories (design, procurement, safety, etc.).
- Identification of causes that demand urgent attention.
- Risks requiring immediate response vs. those for later analysis.
- A **watchlist of low-priority risks** for monitoring.
- A probability vs Risk impact diagram, as shown below, is prepared to prioritise the risks.

Probability vs Impact (Risk Matrix) is depicted in the following figure 1. Probability vs Impact (Risk Matrix). This is also called a heat map. Helps us to prioritise risks.

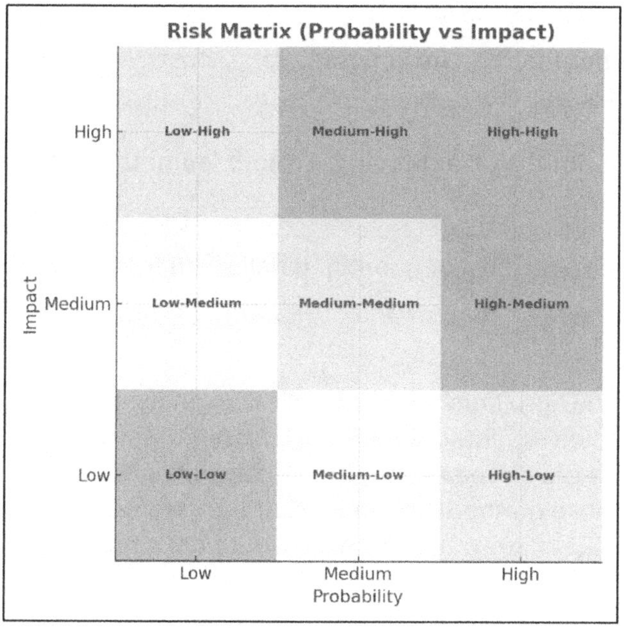

Figure 5.8 Probability vs Impact (Risk Matrix)

Quantitative Analysis:

For **high-priority, high-impact risks**, a more detailed quantitative analysis may be required. This involves **numerical measurement** of risk probability and consequences to estimate their implications on project objectives.

Common tools include:

- **Probability Distributions** – Using methods such as "method of moments" to estimate project range outcomes.
- **Expected Monetary Value (EMV)** – Calculating risk exposure as *Probability* × *Cost*.
- **Three-Point Estimating (PERT)** – Using Pessimistic (L), Most Likely (M), and Optimistic (H) estimates with probabilities for each to calculate expected values.

The **updated Risk Register** after quantitative analysis provides:

- **Probabilistic forecasts** for cost and schedule performance.
- **Probability of achieving project objectives** (time, cost, quality).
- A prioritised list of quantified risks with trends over time.

Practical Note

Most HVAC projects stop at qualitative analysis. Risk registers are typically updated with a prioritisation matrix or heat map, and risks are sorted based on their severity and urgency. While quantitative methods are powerful, they are generally used for large-scale, high-value projects where risks have a significant financial or contractual impact.

Golden Rule – Risk Assessment - *"Not every risk is a high priority. Focus resources on the few critical risks that can make or break the project."*

5.11.9 Risk Response Strategies

Risk response planning is the process of developing options and defining actions to minimise uncertainties and protect project objectives. The aim is not only to reduce negative risks (threats) but also to be prepared for opportunities that may benefit the project.

Characteristics of Effective Risk Responses

To be successful, every risk response should meet the following characteristics:

- **Proportional:** The response must be appropriate to the severity of the risk.
- **Impact-Oriented:** It should minimise the effect of the risk on cost, schedule, or quality.
- **Cost-Effective:** The cost of the response should not outweigh the value of risk reduction.
- **Timely:** Responses must be swift and proactive, rather than reactive.
- **Practical:** The response should be realistic within the project context and available resources.
- **Collaborative:** Responses should be discussed, agreed upon, and accepted by all stakeholders.
- **Owned:** Each response must have a clearly defined owner responsible for implementation.

Risk Response Strategies:

There are multiple ways to plan a risk response. A few of the commonly used responses are discussed below.

- **Risk Avoidance**: Risk avoidance is the best response for a given risk. It eliminates the risk altogether.

For example,

  - Freeze technical submittals early to avoid frequent design changes.
  - Use safety protocols strictly to avoid accidents, loss of life and time at the site.
  - Using **3D BIM modelling** to identify and avoid duct/pipe clashes with structural beams before fabrication.

- **Risk Mitigation**: The Most frequently used risk response is Risk Mitigation. It is done to reduce the probability or impact of the risk through proactive measures. (alternative vendors, buffer in schedule).

Example in HVAC Projects:

  - For imported chillers with long lead times, mitigate risk by placing the PO early and monitoring vendor progress with weekly reports.
  - Stocking 5–10% extra copper piping or insulation to mitigate the risk of minor shortages at the site.
  - Use multiple piping contractors to mitigate the risk of delay due to the absence of contractors

- **Risk Transfer:** Risk transfer is basically shifting the responsibility of the risk to a third party, usually through contracts, insurance, or warranties. (insurance, warranties, subcontract agreements).

Example in HVAC Projects:

- Requiring suppliers to provide performance guarantees and extended warranties for major equipment like chillers and pumps.
- Taking transit insurance for imported AHUs to transfer the risk of damage during shipping.

- **Acceptance:** These are typically external risks or inherent risks for which no mitigation measures are possible. It is better to acknowledge the risk but take no immediate action, instead planning contingencies if it occurs.
    - Accepting the possibility of minor rain delays during duct installation, but building in schedule float.
    - Budgeting a contingency fund for potential fluctuations in steel or copper prices.

5.11.10 Reserves Management

5.11.10.1 Contingency Reserves:

Contingency reserves are **funds or time buffers** built into the project plan to cover known risks **("known unknowns")** that are not included in the baseline estimates. The size of the reserve is determined by the potential impact of identified risks and should be enough to implement contingency actions plus a buffer for minor, unidentified risks.

Examples of HVAC Projects:

- Adding a two-week buffer in the project schedule to account for delays in chiller delivery due to customs clearance.
- Allocating **extra budget to cover the cost of reordering insulation** or duct accessories if site measurements differ from tender BOQs.

5.11.10.2 Management Reserves

Management reserves are additional funds or time allowances set aside for **unknown unknowns** — events that cannot be reasonably predicted during planning. Unlike contingency reserves, using management reserves requires approval and a formal change to the project's cost or schedule baseline.

Examples of HVAC Projects:

- A sudden government regulation banning imports from a certain country
- A flood or fire at a supplier's factory disrupts delivery of AHUs or pumps.

Key Difference:

- **Contingency Reserve → for *known unknowns*** (anticipated risks, planned buffer).
- **Management Reserve → for *unknown unknowns*** (unpredictable events, formal change required).

The following Table 27 depicts the differences between contingency reserve and management reserve.

Table 5.2821 Contingency reserves vs Management reserves

Aspect	Contingency Reserve	Management Reserve
Definition	Buffer for known unknowns (identified risks not in baseline). Time buffer/funds	Buffer for unknown unknowns (unpredictable events). Funds
Purpose	To absorb the impact of risks anticipated during planning.	To absorb the impact of unforeseen risks beyond the planning scope.
Examples (HVAC)	- Extra 2 weeks for customs delay in chiller delivery. - Budget for reordering insulation/duct accessories. - Spare copper piping at site.	- New govt regulation banning refrigerant type. - Flood/fire at the supplier factory. - Pandemic shutting down imports/labour .
Control	Controlled by the Project Manager / PMO; used directly when risk occurs.	Controlled by senior management; requires a formal change request to baseline.
Approval	Usually within project authority limits.	Requires client/management approval and re-baselining.
Visibility in Plan	Included in cost and schedule estimates.	Not included in baseline; kept separate.

Golden Rule – Reserves: *"Contingencies protect against expected risks; management reserves protect against surprises."*

5.11.11 Updating the Risk Register and Project Plan

Once risk response planning is completed, the **Risk Register** must be updated to include:

- The agreed **risk response actions** for each risk.
- The **responsibility matrix** (who owns, who supports, who monitors each risk).
- At the same time, the **Project Plan** should be updated to reflect any expected changes in **schedule, cost, or scope** arising from risk responses.
- Details of **Contingency Reserves** (for known unknowns) and **Management Reserves** (for unknown unknowns) should be consolidated and submitted for **budget provision and approval.**

Finally, it is the responsibility of **every team member** to ensure that risk responses are executed as agreed, and that outcomes are promptly reported to the **Project Manager/PMO** for tracking and escalation if required.

Case Study – Risk Planning Done Right
Project: Metro Rail Station HVAC System – South India
Scenario: The project team identified early that the imported **chiller** had a 28-week lead time. They included this in the risk register and developed a mitigation plan — ordering in advance, identifying a backup local supplier, and planning partial commissioning with temporary AC units.
Outcome:

- No delays in the overall project schedule.
- Client impressed with proactive planning.
- Team avoided costly penalties.

Lesson Learned: Early risk identification + mitigation ensures smooth delivery, even for critical long-lead items.

Risk Register Template (Sample) is attached in the annexure.

5.11.12 Monitoring & Controlling Risks

Risk monitoring is a **continuous review mechanism** to ensure that risks remain under control and that the project stays aligned with its objectives.

At agreed intervals, the **Risk Register** must be updated with:

- The outcomes of previously identified risks.
- Any **new risks** that have emerged.
- The status of risk response plans and their effectiveness.

When new risks are identified, they should be **assessed qualitatively**, and where necessary, **quantitatively**, to determine their probability and impact. If a mitigation plan has failed, it must be immediately reported to the **Project Manager**, who will escalate it to the relevant stakeholders and trigger a **revised risk response plan**. Any resulting impact on the project's **schedule or cost** should be addressed without delay.

High-impact risks should be monitored closely and, if required, **escalated to the steering committee** as per the predefined escalation plan

Practical Tips for Risk Monitoring

- **Update, do not delete:** Never delete risks from the register. Instead, update their probability and impact as they evolve. Risks that are fully overcome should still be retained for reference.
- **Capture Lessons Learned:** Document resolved risks and their handling in the **Lessons Learned Register**.

- **Share Best Practices:** Any successful mitigation strategies should be communicated to other ongoing projects so the entire organisation benefits.

5.11.13 Conclusion

Risk planning is not about eliminating risks — it is about being **prepared to face uncertainty**. A well-maintained **Risk Register** ensures that surprises are minimised, responsibilities are clear, and contingency plans are ready.

In HVAC projects, where delays or failures can ripple across multiple systems, **proactive risk planning** is often the difference between a project that constantly fights fires and one that delivers smoothly. To achieve this, project teams must sometimes **think creatively and out of the box** to mitigate unique risks.

Risk management is not just about protecting budgets and timelines — it's about protecting trust. In HVAC contracting, **clients remember how risks were handled more than the risks themselves**. A project that identifies risks early, communicates transparently, and acts decisively earns long-term credibility.

Reputation is built not on avoiding risks, but on demonstrating control when challenges arise. A contractor who manages risks well is the one who gets called back for the next project.

Golden Rule – Risk Management: *"If risks are not written down, they are not being managed."*

Risk Planning at a Glance

What is Risk?

An uncertain event that may or may not happen, but can impact project cost, schedule, or quality.

- **Why Plan for Risks in HVAC?**

  HVAC projects have multiple touchpoints (design, procurement, execution, commissioning), and delays or failures in one area ripple through the entire system.

- **Key Steps in Risk Planning**

1. **Identify Risks** – Use checklists, brainstorming, and expert input.
2. **Assess Risks** – Qualitative (probability & impact) and Quantitative (cost/schedule modelling).
3. **Prioritise Risks** – Risk matrix or heat map to focus on high-impact risks.
4. **Plan Responses** – Avoid, Mitigate, Transfer, or Accept.
5. **Allocate Reserves** – Contingency (known unknowns) and Management (unknown unknowns).
6. **Monitor & Control** – Update risk register, track effectiveness, escalate if required.
7. **Tools in Annexure:**
   - *Risk Register (Excel)*
   - *Risk Identification Checklist*
   - *Risk Prioritisation Matrix (Heat Map)*
   - *Risk Response Tracker (optional)*

5.12 Communication Planning

"Clear communication is cheaper than rework, faster than firefighting, and stronger than contracts."

5.12.1 Introduction

Communication failures are among the most frequent causes of disputes, delays, and cost overruns in HVAC projects. This often arises from verbal instructions, untracked approvals, missing updates, or informal channels causing gaps between design, bill of quantities (BOQ), and site execution.

A well-crafted communication plan ensures that information flows effectively to the right stakeholders in the right format and at the right intervals, minimising errors and fostering collaboration.

Communication is a deliberate exchange of information between the project team and its stakeholders spanning the entire project lifecycle. It can be categorised into:

- **Formal vs. Informal**: Formal communication is documented and archived; informal may occur spontaneously but should not replace formal records.
- **Written vs. Verbal:** Written communication provides evidence and traceability; verbal communication is suitable for quick clarifications but must be followed up with documentation.
- **Internal vs. External:** Internal involves project team members; external extends to clients, consultants, vendors, regulatory bodies, and other stakeholders.

In Indian HVAC projects, which often involve multiple consultants, clients, contractors, and vendors, disciplined documentation, dissemination, and retrieval of communications are as important as technical correctness.

5.12.2 Communication Management plan (CMP) Process Flow

The Communication Management Plan (CMP) articulates the communication requirements and protocols for the project.

Purpose: Document and manage expectations about who needs what information, when, how, and from whom, ensuring alignment among diverse participants.

The CMP is a living document and must be regularly updated to reflect project dynamics and stakeholder changes.

Process Flow: Inputs → Process → Outputs

Table 5.2922 Process FloW

Inputs	Process	Output
Project Charter and contractual documents. **Stakeholder Register detailing all parties.** **Project Schedule and key milestones.** **Organisational and client communication policies and standards.**	• Identify all stakeholders and map their communication needs in terms of content, frequency, mode, and responsibility. • Define standardised communication formats: e-mails, weekly status reports, dashboards, trackers, and formal MoMs. • Set communication frequency parameters: daily for critical teams, weekly updates for management, and monthly summaries for clients. • Establish a robust escalation matrix to promptly address issues and conflicts. • Implement document control protocols to govern versioning, approvals, and distribution to prevent misinformation.	• Formalised Communication Management Plan document. • RACI matrix (Responsible, Accountable, Consulted, Informed) specifying communication roles. • Reporting templates customised to stakeholder needs. • Defined escalation protocols with contact details and timelines.

5.12.3 Best Practices

- **Meeting Discipline:** Treat every coordination meeting (physical or virtual) with importance. Ensure clear agendas and participation from all relevant stakeholders.

- **Reporting Structure**: Regularly report progress against plans, highlight potential risks and issues, and document action items explicitly.
- **Minutes of Meeting:** Prepare detailed MoMs and circulate within 24 hours post-meeting for validation and accountability.
- **Single Source of Truth:** Use centralised repositories (cloud platforms, DMS, ERP systems) for storing all live drawings, schedules, and documents accessible to all relevant parties.
- **Version Control:** Maintain strict version control to ensure all parties work with the latest approved documents.
- **Early Escalation:** Detect and escalate issues promptly before they escalate into claims or disputes.
- **Communication Etiquette**: Establish clear email naming conventions and frequency limits to prevent information overload.
- **Encourage Feedback:** Promote an open culture where team members can ask questions and raise concerns through official channels.

Case Study: The Cost of Poor Communication

In a high-stakes hospital HVAC project, a subcontractor began duct fabrication using preliminary drawings shared informally via email by the consultant. When final approved drawings with revised dimensions arrived weeks later, approximately 30% of the ductwork required costly scrapping and re-fabrication. This error caused significant project delays and strained financial performance.

Key Takeaway: Absence of formal, documented communication and strict version control can profoundly impact project margins and timeline.

5.12.4 ☑ Do's and ✖ Don'ts

Table 5.30 Dos and Don'ts

☑ Dos	✖ Don'ts
• Document all verbal instructions and approvals via emails or MoMs • Track approvals and decisions using a centralised log. • Use dashboards with summaries (e.g., traffic light system) for reporting to senior management. • Keep messages clear, concise, structured, and actionable. • Tailor communication at a detailed level to audience needs.	• Don't assume WhatsApp or informal messages substitute official communications • Don't expect immediate responses on informal communication channels. • Don't circulate multiple versions of drawings — always send the latest revision. • Don't leave escalations vague — define time limits (e.g., unresolved in 48 hrs → escalate to PMO). • Don't: Mark everyone on every email (info-overload = missed action points).

5.12.5 Conclusion – Communication Planning

Communication is not an accessory to project management; it is the bloodstream that keeps an HVAC project alive. A project with excellent engineering and procurement can still fail if communication breaks down. Misunderstood instructions, untracked approvals, and scattered drawings turn into costly rework, strained relationships, and payment disputes.

A robust communication plan transforms this chaos into control:

- **Clarity**: everyone knows what to do and when.

- **Accountability:** responsibilities are documented and traceable.

- **Transparency:** decisions, approvals, and escalations are visible.

- **Trust**: clients and consultants see professionalism and discipline.

In the Indian HVAC industry, where projects involve multiple consultants, contractors, and vendors, structured communication is often the deciding factor between timely completion and extended arbitration.

> **Golden Rule:** *"A project manager who communicates with discipline doesn't just prevent disputes — they deliver peace of mind to every stakeholder."*

5.13 Integrated Project Planning in HVAC

"In HVAC projects, success is not built in silos — it comes from weaving every plan into one seamless roadmap."

5.13.1 Introduction

It is common in HVAC industry to assume planning means a Gantt chart. It is only a part of the Integrated Planning. Integrated Planning is the art of bringing together every strand of the HVAC project — scope, schedule, cost, resources, risk, quality, procurement, commissioning, and communication — into a single, coherent plan.

In HVAC projects, subsystems cannot function in silos. Ducting depends on civil openings, piping depends on equipment delivery, commissioning depends on timely electrical readiness, and cash flow depends on accurate measurements. A plan that addresses each of these individually but not together will fail at the first site clash.

Integrated Planning ensures that **all disciplines, trades, and stakeholders are aligned** and that **every dependency is visible** before execution begins.

5.13.2 Integrated Project Planning Process flow

Inputs → Process → Outputs

Table 5.31 Integrated Project Planning 23 Process Flow

Inputs	Process	Outputs
• Project charter - Contract BOQ/ specifications • Approved drawings - Risk register - Resource estimates	• Consolidate scope, WBS, and BOQ • Link schedule with procurement & cash flow • Align QA/QC, risk, Resources, and HSE plans • Integrate communication protocols • Simulate scenarios (best case, delays, resource shortages)	• Integrated Project Plan (Master Plan) • Linked Scope (WBS)-schedule-cost-resource baseline • Consolidated risk, procurement, & quality framework • Communication matrix & escalation path

5.13.3 Best Practices in Integrated Planning

- Always link **procurement, schedule, and cash flow** — never plan them in isolation.
- Prepare a **single integrated tracker** covering drawings, submittals, procurement, schedule, quality, and billing.
- **Integrate activities of other trades** that are directly related to HVAC. For example, wall openings, false ceilings, Plan room readiness and so on.

- Review integration at **weekly project reviews** — scope changes, delays, and site issues must reflect across cost, schedule, and resources immediately.
- Include **commissioning early** — don't treat it as the last stage; link its prerequisites into the main schedule.
- Use **visual dashboards** to present integrated status (traffic light for cost, schedule, risk).

Project Management effectiveness is in anticipating what could go wrong and what would potentially delay HVAC deliverables, and taking preventive measures.

5.13.4 Do's and Don'ts – Integrated Project Planning

Table 5.32 Do's & Don'ts Integrated Project Planning 24

✅ Dos	❌ Don'ts
• Create a master integrated plan before mobilisation. • Involve all stakeholders (consultant, client, vendors) in planning workshops. • Regularly update the plan and ensure downstream trackers are aligned. • Use digital tools (MS Project + Excel dashboards + ERP) for integration.	• Don't allow procurement and engineering to plan in isolation. • Don't keep multiple versions of the "master plan" circulating. • Don't ignore risk and cash flow while focusing only on the schedule. • Don't treat commissioning as a last-minute activity.

5.14 Conclusion – Project Planning Core for HVAC

Project planning is the bedrock upon which the success of any HVAC project rests. Throughout Part II – Planning Core, we have examined the foundational elements that transform ideas and designs into practical, deliverable realities within the HVAC domain. Each subtopic—ranging from Work Breakdown

Structures to Integration Management—serves as a critical gear in the engine that drives project delivery.

5.14.1 The Power of Structured Planning

A robust project planning phase sets expectations, organises resources, controls timelines, and pre-empts challenges. The journey begins with a clear **Project Planning Overview**—establishing objectives, deliverables, and the project roadmap. This big-picture view aligns the client, consultants, and executing teams, making transparent the "what," "why," and "how" of project realisation.

Breaking down the project into manageable components through the use of a **Work Breakdown Structure (WBS)** ensures that nothing is missed and responsibilities are allocated appropriately. Every activity—whether engineering, procurement, installation, or commissioning—is tracked, preventing scope gaps and helping teams anticipate resource and timeline needs.

5.14.2 Technical Rigour and Documentation

Documentation is not just paperwork: it is the language of coordination and accountability. **Drawings and Shop Drawing Approvals** guarantee that all parties build from the same design intelligence, minimising ambiguity and reducing the risk of costly rework. **Technical Submittals** reinforce this by formalising material and product selections, ensuring conformance to specifications.

Method Statements serve as tactical guides—outlining step-by-step execution, safety requirements, and quality controls for each major activity. Such statements become invaluable references for both site teams and supervisors, aligning operations with best practices and client expectations.

5.14.3 The Art of Scheduling and Controls

No project progresses without time discipline. **Scheduling** in HVAC projects often involves dependencies between trades, long-lead procurement items, and tight windows for installation. Developing a realistic, resource-loaded schedule is more than a Gantt chart—it is a living tool for monitoring progress, predicting constraints, and responding to unforeseen changes with agility.

Closely linked are **Cost Management** and **Resource Management**. Professional HVAC planners are adept at preparing accurate budgets, tracking expenditures, and reallocating resources to prevent downtime. Cost overruns, often rooted in reactive procurement or ill-timed mobilisation, are best handled through proactive planning and vigilant management.

5.14.4 Procurement, Quality, and Risk: Balancing Agility and Assurance

Procurement Management demands a blend of technical understanding and commercial foresight. Timely procurement of compliant materials often defines the critical path in HVAC projects, with global supply-chain challenges amplifying the need for early, well-documented purchasing.

Quality Management is never an afterthought. From design through handover, robust quality systems ensure the installed HVAC systems perform reliably, comply with codes, and deliver on sustainability promises. Quality is tracked through systematic inspections, test plans, and commissioning protocols.

Meanwhile, **Risk Management** requires professionals to anticipate (and mitigate) what could go wrong, ranging from technical hitches and site conditions to vendor and regulatory uncertainties. Formal risk identification, prioritisation, and action planning—documented and reviewed regularly—inoculate projects against costly surprises.

5.14.5 Communication and Integration: The Glue of Project Planning

Even the most detailed plan can unravel if communication falters. **Communication Management** stands at the heart of effective coordination, ensuring all team members, from subcontractors to clients, receive the right information, at the right time, in the right format. Centralised repositories, standardised reporting templates, and disciplined meeting structures eliminate ambiguity while building trust.

Finally, **Integration Management** acts as the orchestrator, bringing together the myriad strands—drawings, method statements, procurement, resources, quality, risk, and communications—into a unified plan. The best planners visualise the interdependencies, adapting as project realities evolve while keeping all stakeholders aligned on updated objectives.

5.14.6 The Path Forward

HVAC project planning is neither a static nor isolated task—it is an ongoing process of review and realignment. Successful professionals treat the planning core as both a toolkit and a mindset, blending technical rigour with flexibility, collaboration with leadership, and control with innovation.

A well-rounded project planner is not one who merely completes schedules and checklists, but who anticipates challenges, communicates clearly, manages resources efficiently, and ensures that each stage—no matter how routine—is executed as a foundation for project success. The principles detailed across this Planning Core serve as proven strategies to steer projects away from chaos and toward excellence.

In the end, the measure of good project planning is not the absence of problems, but the preparedness to handle them—delivering HVAC systems that perform, endure, and exceed expectations.

Let these core principles and best practices be your guide as you progress to the subsequent phases of project execution and control. Great planning does not guarantee a perfect project, but it will always create the strongest foundation for achieving it.

Chapter 6: Project Execution

"Execution is where plans meet reality — and reality rewards only those who prepare for it"

Learning Objectives

By the end of this chapter, the reader will be able to:

1. Translate **approved plans** (drawings, submittals, method statements) into site **execution**.
2. Prepare the **site and ensure readiness** (civil handover, scaffolding, safety, access).
3. **Implement HSE protocols** during all installation activities.
4. **Manage procurement, dispatch**, and site storage efficiently.
5. Execute installation of **major HVAC systems** (air distribution, chilled water piping, plant room, VRF, electrical panels, controls).
6. Identify **risks during execution** and apply corrective measures.
7. Use checklists, trackers, and QA/QC protocols for quality execution.

6.1 Introduction

Project Execution is the stage where all the planning, approvals, and preparation translate into action at the site. In HVAC projects, execution is the most resource-intensive and risk-prone phase. While **Initiation** sets the foundation and **Planning** defines the roadmap, it is in **Execution** that project managers prove their leadership, foresight, and ability to deliver.

Execution involves translating approved drawings, technical submittals, and method statements into actual installation, with quality, safety, and cash flow closely monitored. The execution phase is where designs transform into actual ducts, chillers, pipes, and plant rooms. It is also the phase where most project risks materialise — delays in civil works, wrong materials at the site, untrained manpower, unsafe practices, and poor coordination across trades.

While planning provides the roadmap, execution is about **driving on that road** with speed, safety, and precision. In the Indian HVAC industry, where projects are often fast-tracked and dependent on multiple subcontractors, execution discipline makes the difference between **smooth delivery and costly rework**.

A common observation: *Projects that looked perfect on paper during planning often fail because execution teams bypass approvals, ignore site readiness, or rush activities under client pressure. The result? Idle labour, material wastage, safety incidents, and commissioning stress.*

Well-executed projects, however, are marked by:

- **Early identification of bottlenecks.**
- **Strong documentation and approvals before starting work.**
- **HSE culture integrated into daily activity.**
- **Proactive coordination with civil, electrical, and interiors.**
- **Quality-first mindset, ensuring every stage is verified before moving ahead.**

The Project Manager's role during execution is to transform the paper plan into a working system — on time, within budget, and to specification.

> **Golden Rule:** *"Execution is not about speed alone — it is about safe, approved, and coordinated progress."*

6.2 Execution Framework Narrative

To ensure systematic delivery, the Execution Framework for HVAC projects can be viewed as a step-by-step process as given below:

Step 1: Document Submission & Approvals

- Shop drawings, technical submittals, and method statements are finalised and approved before execution starts.
- Only the latest approved version is circulated to site teams and subcontractors.

Step 2: Site Preparation & Readiness

- In the meantime, the site execution team visits the site to understand the site's readiness for project execution. Civil handovers, scaffolding, floor openings, utilities (power, water), and access must be checked.

- HSE protocols and permits-to-work must be in place before mobilising manpower.

Step 3: Health, Safety & Environment (HSE)

- Safety inductions, toolbox talks, PPE enforcement, and safe working practices.

- Integration of HSE into daily operations rather than treating it as compliance paperwork.

Step 4: Procurement & Material Dispatch

- Link all dispatches to approved BOQ deviations and site readiness.

- Sensitive materials (copper pipes, insulation, GI sheets) are dispatched only when installation is ready to avoid damage and theft.

Step 5: Installation Activities (Construction Activities)

- **Air Distribution**: ducting, insulation, grills, and diffusers.

- **Hydronic Systems**: chilled water piping, condenser water piping, pumps.

- **Plant Room Works:** chillers, cooling towers, foundations, and accessories.
- **VRF Systems**: indoor and outdoor units, refrigerant piping, leak testing.
- **Electrical & Controls**: panels, cabling, IBMS, IoT devices.
- **AHUs/FCUs/TFAUs**: installation, alignment, and connection.

Step 6: Measurements & Joint Records

- **Continuous measurement of quantities (ducting, piping, insulation) and joint measurement records with PMC/consultant.**
- Prevents disputes later during billing and closeout.

Step 7: Pre-Commissioning

- Flushing, pressure testing, cleaning, and functional safety checks before commissioning.
- Deficiencies logged in snag lists for closure.

Step 8: Integration with Commissioning

- Execution team must hand over a system that is ready for commissioning — complete, safe, inspected, and documented.

This framework ensures execution does not become a fire-fighting exercise but remains a structured, verifiable, and safe journey from planning to commissioning.

6.3 Detailed Installation (Construction) Process

6.3.1 Document Submission & Approvals

a. Shop Drawings

During planning, shop drawings were developed, reviewed, and approved.

In execution:

- Ensure only the latest approved drawings are used at the site. Maintain a Drawing Control Register.
- It is the responsibility of the site engineer to validate the dimensions as per site conditions and raise a red flag in case of any deviation, even before the erection starts.
- Any minor/ major changes that happen at the site MUST be informed to the design team immediately. An impact analysis must be carried out.
- Print and display approved zone drawings in site cabins.
- Conduct weekly drawing coordination meetings with civil and electrical contractors.
- Avoid shortcuts **X:** never install ducts/pipes based on "**oral approvals.**"

Best Practice Zone-wise drawings should be released for execution in sync with the schedule — not in bulk. This avoids site congestion and confusion.

b. Technical Submittals

- Submit manufacturer datasheets, compliance sheets, and test certificates before ordering.

- Link each submittal to the drawing number and BOQ item.
- Approved submittals now guide material procurement and installation.
- Cross-check deliveries against approved submittals (model, capacity, materials, construction, finish).
- Attach submittal references to Material Inspection Reports (MIRs) at the site.

Any deviation from the material should trigger a site NCR (Non-Conformance Report). Any deviation should be reported to the concerned Authority immediately.

- Track approvals via a **Technical Submission Status Tracker.**

c. Method Statements

Each method statement created during project planning must now be implemented:

- Site engineers must be trained to read the method statements correctly and implement the processes as described therein.
- Supervisors should keep printed method statements for **ducting, piping, insulation**, and all installation (construction activities), etc.
- Daily toolbox talks should cover method + HSE points.
- ITP (Inspection & Test Plan) checklists must be signed daily by QA/QC.

A sample method statement is shown in the annexure

6.3.2 Site Preparation and Readiness

Before starting any installation activity, **site readiness** must be validated. Too many projects fail because materials arrive at the site **before civil handovers, scaffolding, or utilities are ready**. The result: **idle manpower, wasted material, and strained client relations.**

Key Checks Before Mobilisation

1. **Civil Readiness**

- Verify wall openings, shafts, and floor cutouts are provided as per approved drawings.
- Plant rooms and AHU rooms must be free from debris, with floors levelled and access doors operational.
- False ceiling levels frozen before duct installation starts.

2. **Worker & Staff Accommodation in Major Projects**

In large-scale HVAC projects, where engineering and supervision staff are required to stay within the site campus for extended periods, proper living arrangements are not just a welfare measure — they are a project enabler. Poor conditions directly affect morale, productivity, and even retention of skilled manpower.

Key Considerations:

- **Safe & Habitable Sheds/Rooms:** Ensure temporary accommodation is constructed with proper ventilation, adequate space, and protection against weather.

- **Power & Lighting:** Provide reliable and safe electrical connections with proper earthing and distribution boards. Avoid unsafe cabling or makeshift arrangements.

- **Water & Sanitation:** Drinking water facilities and hygienic toilets are essential. Lack of these leads to absenteeism, health issues, and regulatory non-compliance.

- **Safety from Wildlife & Hazards:** In remote sites, sheds should be designed to keep away snakes, rodents, and other wild animals. This includes raised flooring, proper sealing, and controlled access points.

- **Basic Amenities:** Fans, cots, and mosquito protection make a big difference in comfort and worker satisfaction.

- **Fire & Emergency Preparedness:** Each accommodation block should have fire extinguishers, emergency exits, and periodic safety checks.

> **Golden Rule:** *"Healthy, safe, and comfortable workers deliver better projects. Accommodation is not an overhead cost — it is an investment in productivity.*

3. **Utilities**

- Permanent or temporary power supply available for tools, welding machines, and lifts.

- Use proper electrical connections to draw power from the common source. Avoid dangerous and unsafe methods, such as putting wires without a switch to draw power.

- Ensure Water supply available for testing, flushing, and cleaning.

- Adequate lighting for safe working conditions.

4. **Access & Scaffolding**

- Safe scaffolding or aerial platforms erected and certified before high-level works.
- Mobile scaffolding equipment is available at site and is fully functional and tested for its safety.
- Lifts/hoists available for heavy equipment (chillers, AHUs, cooling towers).
- Clear pathways for material movement without obstruction.

5. **Coordination with Other Trades**

- Confirm with civil, electrical, firefighting, and interiors on work sequence.
- Resolve space conflicts in shafts and corridors before mobilisation.

6. **Documentation**

- Approved shop drawings and method statements are available at the site.
- Latest revisions displayed at the site office (no outdated versions).
- Kick-off meeting held with client/consultant before starting execution.

Golden Rule: *"Never send materials to the site unless the site is certified ready.*

6.3.3: Health, Safety & Environment (HSE)

Safety is often compromised under pressure to "**get the work done quickly.**" Yet, every unsafe act risks not only the project but also the lives of those involved. In HVAC projects, where activities involve heights, heavy lifting, welding, confined spaces, and electrical work, **HSE integration is non-negotiable.**

Core HSE Practices for HVAC Execution

1. **Induction & Training**

- Every worker must **undergo site induction** before entering the site.
- Toolbox talks are conducted daily before the shift starts.
- Special training for high-risk jobs: hot works, working at heights, confined spaces.

2. **Permits-to-Work (PTW)**

- Hot Work Permit for welding, cutting, and grinding.
- Height Work Permit for ducting/piping above 2 m.
- Confined Space Permit for shafts, tanks, and plant rooms.

3. **Personal Protective Equipment (PPE)**

- **Mandatory helmets, safety shoes, reflective jackets, gloves, and goggles.**

- Harnesses for working at heights.
- Earplugs and face shields for noisy or spark-producing activities.

4. **Emergency Preparedness**

- First-aid kits and fire extinguishers available at all work zones.
- Details of medical teams' availability and emergency contact numbers to be available at the site.
- Emergency exit routes marked and illuminated.
- Incident reporting system in place (near misses, accidents, unsafe observations).

5. **Environmental Protection**

- Proper storage of chemicals, adhesives, and refrigerants.
- Waste segregation and disposal as per local norms.
- Noise and dust control measures, especially in occupied buildings.

6. **Monitoring and Reporting**

- **Dedicated Safety Officers** on site.
- Regular **mock drills** (fire, fall rescue).
- Mandatory PPE usage enforced by safety officers.
- **Toolbox talks** daily before work.

- Daily toolbox talks and weekly safety audits.
- Proper scaffolding and lifting equipment.
- Prominently display "accident-free days" at the site.

Golden Rule: *"Flawless execution is not about working harder — it is about working safer, smarter, and together."*

HSE Case Study

At a commercial office project in Pune, duct installation was started without scaffolding certification. A worker slipped while fixing ducts above 4 meters, resulting in serious injury. Work was halted for two weeks for investigation, delaying not only ducting but also false ceiling and electrical work. The worker was unable to work later.

Lesson: Rushing into work without certified scaffolding or permits leads to accidents, stoppages, and reputational damage. It can cause irreparable damage to a family.

6.3.4: Procurement and Material Dispatch

- Procurement teams should release material dispatches in sync with the **baseline schedule**.
- All received materials should undergo **MIR inspection**.
- Maintain a **site material log** (quantities received vs. BOQ vs. approved deviations).
- Return protocols: if the project doesn't allow early returns, excess stock should be accounted for separately to avoid "ghost stock."

Procurement & Material Dispatches

Procurement in the planning stage defines **what, when, and how much** needs to be ordered. But in the execution stage, procurement success depends on **accurate dispatching, site readiness, and material control.**

Too many HVAC projects in India face losses not because of wrong BOQs, but because materials are **sent too early, stored improperly, or not reconciled with approved drawings.**

6.3.4.1 Material Dispatch Workflow

A. Approval Gate

o **Technical Submittals**: No procurement unless the submittal is approved.

- **Deviation Statement**: Only dispatch quantities approved against BOQ deviations.

- **Site Readiness Confirmation**: Send only when foundations, scaffolding, and access are ready.

B. Dispatch Checklist (per material type)

- **GI Sheets/ Fabricated Ducts**: Dispatch after duct routing approval and the site is ready for erection

- **Copper Pipes**: Send only when piping routes are marked and wall openings are ready.

- **MS Pipes for Chilled water piping and Condenser water piping**
- **Valves & Accessories**: Send in lots linked to the installation sequence, not bulk.

- **Insulation Material**: Send in phases; avoid dumping entire stock to prevent damage and loss of materials

Lesson: Never purchase materials without a proper Purchase Order (PO) and delivery note.

In the rush to meet urgent site requirements, teams sometimes bypass the procurement process. While this may seem like a quick fix, it creates two serious risks:

- **Unaccounted Costs:** Materials procured without documentation often don't get logged into project records, leading to cost overruns that cannot be justified during audits or client billing.

- **Quality Risks:** Unverified materials, procured outside the approved vendor list, may not meet specifications. This results in rework, failures, or disputes during commissioning.

Best Practice:

- Always route purchases through the approved procurement system.

- If urgent, raise an emergency PO or at least obtain written client/consultant approval before ordering.

- Ensure every delivery note is signed, logged, and linked to BOQ/variation orders.

C. Documentation

- Delivery Challan with material details, BOQ ref no., and site location.

- Material Inspection Report (MIR) to be signed at the site during receipt.
- Goods Received Note for all the materials received and accepted against a PO should be maintained.
- Maintain a Material Dispatch Tracker linked to BOQ quantities.

Lesson: Get the material utilised (measured by the customer preferably) zone-wise regularly, and it should be validated before dispatching any additional quantity.

D. Storage at Site

- Covered and secured storage for copper, insulation, and electrical items.
- Proper racks for GI sheets and ducts (avoid bending/denting).
- Identify responsibility — **who owns material security at the site?**

Lesson: Despite the best security measures, materials are still lost at the site. MS Pipes, accessories are used by other trades such as fire, plumbing. People may mistakenly use the information available openly at the site.

Best Practices in Procurement Execution

- **Just-in-Time Dispatch**: Don't flood the site with materials. Dispatch as per work-front readiness.

- **Link Dispatch to WBS**: Align every dispatch with schedule activity IDs.
- **Secure High-Value Materials**: Copper, controls, and VRF units require locked, monitored storage.
- **Return Policy**: Agree contractually if unused material can be returned or transferred.
- **Documentation Discipline**: Every movement recorded → BOQ ref → site entry → inspection.

Golden Rule: *"Material sent to the site without readiness is not inventory — it's liability."*

☑ Do's and ✗ Don'ts of Procurement and Material Dispatch

Table 6.1 Do's & Don'ts Procurement and Material Dispatch

☑ Dos	✗ Don'ts
Dispatch materials only after **shop drawings and submittals are approved.**	Don't send GI sheets or copper pipes to site before drawings are frozen.
*Align dispatches with **work-front readiness*** (duct openings, scaffolding, foundations completed).	Don't dump insulation or accessories at site months before use — they get damaged.
Use the **Material Dispatch & Reconciliation Tracker** linked to BOQ.	Don't send quantities beyond BOQ without written client/consultant amendment.
Ensure **covered, secure storage** for copper, insulation, controls, and electricals.	Don't leave sensitive materials exposed to dust, rain, or theft.
Dispatch in **phased lots** based on WBS and installation sequence.	Don't supply in bulk to "clear stock" — it becomes site liability.
Get a **signed MIR (Material Inspection Report) and Goods Received Note (GRN)** at receipt to avoid disputes.	Don't buy any materials without a PO. Don't dispatch without a delivery challan. Don't rely on verbal confirmations — undocumented receipts create payment disputes.

Case Example – Good Practices

At a large hospital project in Chennai, the contractor created a **Material Dispatch Matrix** linking every BOQ item to installation zones.

- Copper was sent floor-by-floor only when shaft openings were complete.
- Insulation was sent in weekly lots.
- A daily **Material Reconciliation Sheet** ensured that installed vs. issued vs. BOQ was tracked.

Result:

- Minimum wastage of copper and insulation.
- No rejections by the consultant.
- Project achieved 2% savings on material cost — purely from disciplined dispatching.

6.3.5: Installing (construction) Major Systems

6.3.5.1 Air Distribution System

Introduction

Air distribution is the backbone of HVAC projects. Chillers, pumps, and AHUs generate and move energy, but it is the duct network and air products — diffusers, grilles, dampers — that actually deliver comfort to the occupants. If the air is not distributed properly, even the best-designed plant will fail to satisfy end users.

A common site mistake is inadequate or poorly planned return air passages. This results in a circulation imbalance, discomfort for occupants, and often costly rework to correct the issue. Properly designed and executed air distribution ensures not just thermal comfort, but also energy efficiency and system reliability.

In execution, ducting installation is one of the **most coordination-intensive activities**. It intersects with civil works (shafts, wall openings), electrical trays, fire-fighting pipes, and false ceilings. A single error in routing or sequence can result in major rework, delays, and disputes.

> **Golden Rule**: *"A duct wrongly routed during execution is not just a mistake — it is a chain reaction that can derail ceiling works, electrical cabling, and interiors."*

Figure 6.1 Typical Air distribution from AHU

Execution Framework for Air Distribution

Pre-Requisites

- Approved shop drawings (signed by consultant/client).
- Deviation statement approved (BOQ vs site quantities).
- Civil readiness: shafts open, beams verified, ceiling heights checked.
- Storage: Ore fabricated ducts and GI sheets are protected from rust and moisture.

1. Duct Fabrication

- Fabrication as per SMACNA/IS standards.
- Correct gauge and thickness (validated via Vernier/micrometre).
- Reinforcement (tie rods, stiffeners) as per standard tables.
- Sealant application to avoid leakage at joints.

- Proper labelling of ducts before dispatch to site.

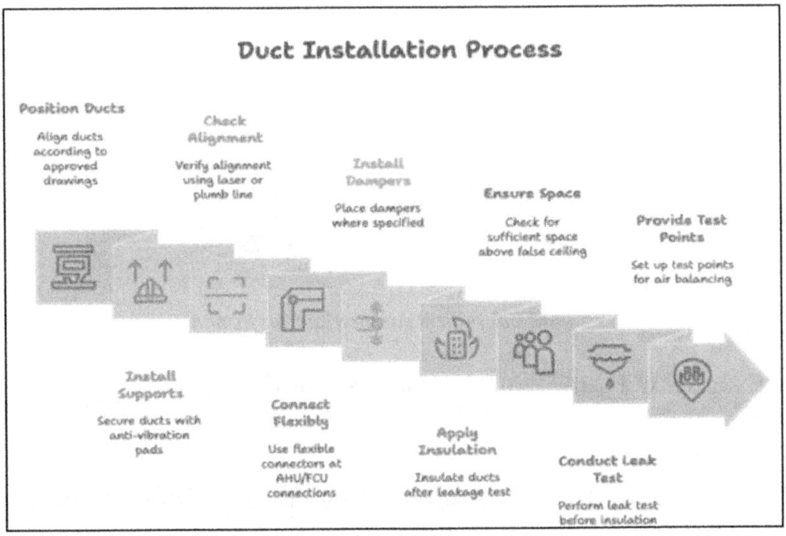

Figure 6.2 Duct Installation Process

2. **Duct Installation**

- Positioning as per approved shop drawings and coordinated sections.
- Supports installed with anti-vibration pads.
- Alignment checked with laser/plumb line.
- Flexible connectors used at AHU/FCU connections.
- Dampers installed where specified (fire dampers, VCDs).
- Insulation applied after leakage test.

- If return air is collected above the false ceiling, ensure there is enough space, and the passage is clear without any obstructions.
- Conduct a duct leak test before insulating. Ensure the joints are closed for minimum acceptable leakage.
- Provide necessary test points for Testing and air balancing.

2.1 Test Points on Ducts for Air Balancing

Air balancing ensures that the designed air quantities are delivered at each diffuser, grille, and duct branch. To make this possible, **test points** must be provided in the ductwork at strategic locations.

Where to provide test points:

- **Main Supply Ducts** – one test point before and after each major branch. The number of points will be decided based on the TAB plan.
- **Branch Ducts** – at least one test point per branch, preferably close to the balancing damper.
- **Before and After Equipment** – provide test points upstream and downstream of AHUs, FCUs, and fans.
- **Across Filters and Coils** – for measuring pressure drop (ensures cleanliness and performance).
- **At Fire Dampers & VCDs** – to check static pressure and ensure proper damper setting.

- **Return Air Ducts** – provide at least one per floor/zone to measure return air flow and static.

- **Exhaust Ducts** – especially for kitchen, toilet, and lab exhaust, where balancing is critical.

2.2 Design & Installation Notes:

- Use **rubber grommets** or **test ports** (with plastic/metal plugs) to allow insertion of Pitot tubes or probes without leakage.

- Ensure ports are **accessible** (avoid false ceiling obstructions).

- Clearly **label test points** (with tags or paint marking) for future balancing and maintenance.

- Recommended size: 10–12 mm diameter hole with a tight-fitting removable cap.

> **Golden Rule:** *"If you can't measure it, you can't balance it. Always provide test points in ducts during installation — retrofitting later is messy, inaccurate, and expensive."*

3. Air Products (Diffusers, Grilles, Louvres)

- Installed only after duct alignment and false ceiling cut-outs are finalised. Use flexible connections where required.

- Ensure grills/diffusers are fitting properly into the duct without any leakage.

- Level and alignment checked with ceiling finishes.

- **Volume Control Dampers (VCDs)** should always be installed at appropriate locations to regulate airflow and ensure proper balancing across the duct network. Position them where they are easily accessible for operation and routine maintenance.
- Inaccessible or poorly placed dampers often result in uneven air distribution and make future adjustments or service extremely difficult. Balancing dampers fitted behind diffusers where required.
- Fire dampers must be installed at all **fire-rated partitions, walls, and floors** as per NBC/ASHRAE/ISHRAE and local fire authority standards.
- Ensure finish matches architectural requirements (powder coating, anodising).

4. Integration with Other Systems

- Weekly coordination with civil, interiors, electrical, and plumbing trades.
- Ensure no clash with cable trays, fire sprinklers, or lighting fixtures.
- Maintain minimum clearance distances for serviceability.

☑ Do's & ✗ Don'ts of Air Distribution Installation

Table 6.2 Do's & Don'ts Air Distribution Installation

☑ Dos	✗ Don'ts
• Always cross-check duct sizes against shop drawings before fabrication. • Protect fabricated ducts with plastic wrap or tarpaulin until installation. • Use laser alignment tools for duct levelling. • Ensure adequate access doors are provided for cleaning/maintenance. • Perform leakage tests before applying insulation.	• Don't proceed with fabrication without final approved drawings. • Don't install ducts in areas where ceiling/wall openings are not ready. • Don't leave ducts unprotected at the site — rust and dents cause rework. • Don't bypass dampers or leave them inaccessible. • Don't treat diffusers/grilles as decorative items — they are functional components for airflow balance.

6.3.5.2 Chilled Water Piping

Execution Framework for Chilled Water Piping

1. Shop Drawings and Coordination

- Verify coordinated drawings for routing, pipe sizing, and support systems.
- Ensure a detailed piping schematic drawing with sizes and valves, and accessories is available at the site.
- Follow the pipe connections

- Ensure clash detection with other services (electrical trays, firefighting, plumbing) using BIM or overlay drawings.

2. **Material Inspection & Storage**

- Check pipe thickness, material grade, and coatings against technical submittals.
- Store pipes on raised supports to prevent corrosion and contamination.
- Cover both ends of the open pipe while storing.
- Inspect valves, fittings, strainers, and flanges for defects and certificates.

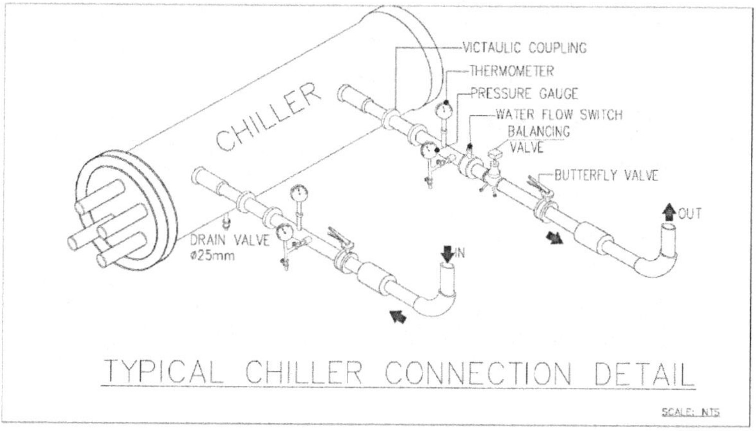

Figure 6.3 Typical Chiller Connection Details

3. **Site Preparation & Readiness**

- Ensure civil openings, trenches, and supports are ready before mobilisation.

- Confirm scaffolding and safe access for higher levels.
- Provide adequate space for pipe expansion, maintenance, and valves.

4. **Pipe Installation**

Pipe installation is one of the most critical activities in HVAC projects since chilled water and condenser water networks form the lifeline of the system. A small oversight during installation can result in leaks, pressure drops, or access problems during commissioning and operation.

Key Guidelines

- **Follow Method Statements**
  - Adhere strictly to the approved method statements for cutting, grooving, welding, threading, or jointing.
  - Ensure WPS (Welding Procedure Specification) and approved techniques are followed.
- **Pipe Cleaning**
  - Clean both external and internal pipe surfaces before welding or jointing.
  - Remove rust, scale, oil, and debris to prevent contamination and corrosion.
- **Supports & Hangers**
  - Provide adequate supports, hangers, anchors, and vibration isolators at recommended intervals.
  - Verify alignment and spacing as per IS/ASHRAE standards.

- **Slope & Drainage**
  - Maintain appropriate slope in pipelines for air venting and proper drainage.
  - Install air vents at high points and drain valves at low points.
- **Accessibility**
  - Valves, strainers, thermometers, and pressure gauges must be installed at accessible locations for easy operation and maintenance.
  - Allow space for future replacement and repairs.
- **Test Points**
  - Provide adequate test points for flushing, pressure testing, and balancing.
  - Ensure test ports are marked and documented in the drawings for future reference

Golden Rule – Pipe Installation: *"Pipes carry the lifeblood of the HVAC system. A clean, well-supported, and accessible installation ensures efficiency, safety, and maintainability for years to come."*

5. **Welding & Jointing**

   - Follow qualified welding procedures and use certified welders.

- Follow the Method statements, use approved welding rods.
- Clean pipe edges and remove rust, oil, paint, or moisture.

- Perform visual inspection and NDT where specified.
- Avoid misalignment, poor root penetration, and slag inclusions.

6. **Pressure Testing & Flushing**

- Conduct hydrostatic pressure tests as per standards (IS/ASHRAE).
- Flush the system thoroughly to remove debris, oil, and welding slag.
- Chemical cleaning and passivation to prevent corrosion.

7. **Insulation & Cladding**

- Apply insulation (NBR, PUF, or mineral wool as per design) with vapour barriers.
- Seal all joints and terminations to avoid condensation.
- Provide aluminium/SS cladding for outdoor or exposed piping as per the contract.

8. **HSE Considerations**

- Use proper PPE during welding, cutting, and handling heavy pipes.
- Follow hot work permit system for welding/grinding.
- Ensure scaffolds are inspected and certified.
- Store flammable chemicals (cleaning agents, adhesives) safely.

Table 6.3 Do's & Don'ts

☑ Dos	✕ Don'ts
• Verify pipe material, thickness, and dimensions before installation. • Ensure expansion loops or joints are provided as per design. • Always conduct pressure testing before insulation. • Label supply and return lines for easy identification. • Maintain detailed installation and testing records.	• Don't install pipes in unapproved or clashing routes. • Don't weld without proper electrode storage and procedure qualification. • Don't start insulation without clearance of the hydro test. • Don't block access to valves, strainers, or pumps. • Don't ignore minor leaks — they worsen under operating pressure.

Case Study

Project: IT Campus, Pune

- **Issue:** Contractor insulated the chilled water lines before hydro testing due to schedule pressure.
- **Consequence:** Multiple joints leaked after insulation, forcing re-insulation of entire sections.
- **Impact:** Cost overrun of ₹20 lakhs and delay of 3 weeks.
- **Lesson:** Pressure testing and flushing must always be completed and cleared before insulation.

Golden Rules

- "Hydro test before insulation — always."
- "Access today saves maintenance tomorrow — never block valves or strainers."

6.3.5.3 Insulation

- Thermal and acoustic insulation, fire-rated material compliance.

Insulation is not decoration — it is performance. Done wrong, it costs energy, comfort, and credibility.

1. **Introduction Insulation in HVAC systems plays two key roles:**

Thermal insulation – to minimise heat gain/loss, maintain chilled water and air temperatures, and prevent condensation.

Acoustic insulation – to reduce noise from fans, ducts, and mechanical equipment, ensuring occupant comfort.

Poorly executed insulation leads to condensation drips, high energy bills, corrosion under insulation (CUI), or noise complaints from occupants.

2. **Types of Insulation in HVAC**

a) **Duct Insulation**

- Thermal insulation prevents temperature losses and condensation on supply air ducts.

- Acoustic insulation reduces fan noise and transmission through ducts (especially return ducts).

b) **Pipe Insulation**

- Chilled water, condenser water, and refrigerant piping must be insulated to prevent heat gain and condensation.
- Hot water pipes (if present) are insulated to minimise heat loss.

c) **Plant Room Insulation**

- Chiller plant rooms and AHU rooms require acoustic insulation (panels, boards, or spray foam) to contain equipment noise.
- Thermal insulation is used around large headers, chilled water storage, and buffer tanks.

3. Insulation Materials

Nitrile Rubber (flexible, closed-cell)

- Widely used for chilled water pipes and ducts.
- Good resistance to water vapour permeation.

Fibreglass / Mineral Wool

- Used for both thermal and acoustic applications.
- Requires vapour barriers to avoid condensation.

EPS (Expanded Polystyrene) / XPS

- Used in large-diameter pipe insulation, cold storage, and duct applications.
- PUF (Polyurethane Foam)
- High thermal resistance, used in specialised cold applications.

☑ Do's and ✖ Don'ts of HVAC Insulation

Table 6.4 Do's & Don'ts - HVAC Insulation

☑ **Dos**	✖ **Don'ts**
• Use only approved insulation materials with proper density and thickness. • Ensure surfaces are dry and clean before applying insulation. • Seal all joints and seams with proper adhesives or vapor barriers. • Provide aluminium cladding or jacketing for outdoor and plant room pipes/ducts. • Use acoustic insulation near AHUs, fans, and plant rooms. • Test insulated ducts/pipes for condensation before handover.	• Don't insulate wet or oily surfaces — adhesion will fail. • Don't leave insulation exposed to weather or mechanical damage. • Don't ignore vapor barriers — without them, condensation will penetrate insulation. • Don't use the same insulation type everywhere — thermal and acoustic needs differ. • Don't cover pipes before hydrostatic testing is completed.

6.3.5.4 AHU / FCU / TFAU Installation

"An air-handling unit installed wrong is a lifetime of noise, leakage, and inefficiency. Precision at installation guarantees performance at operation."

1. Introduction

Air Handling Units (AHUs), Fan Coil Units (FCUs), and Treated Fresh Air Units (TFAUs) form the backbone of an HVAC system's air distribution. Their correct installation is critical to achieve the design intent for airflow, temperature control, energy efficiency, and indoor comfort.

Errors during installation — such as misaligned connections, poor vibration isolation, or lack of service space — lead to operational inefficiency, high maintenance, and noise issues.

2. Pre-Installation Readiness

Before installation begins, ensure the following:

- **Civil Readiness**: Foundations or equipment pads are cast, finished, and level.
- **Vibration Isolators**: Spring mounts, inertia bases, or pads are installed as per approved method statements.
- **Access & Service Clearances**: Minimum service spaces provided around AHUs and FCUs for filter replacement, coil cleaning, and motor maintenance.
- **Electrical Power & Drains**: Cable trays, electrical panels, and floor drains ready at equipment locations.

- **Environmental Conditions:** Avoid installation in dusty, unfinished spaces without proper covering.

4. Installation Procedure Air Handling Units (AHUs)

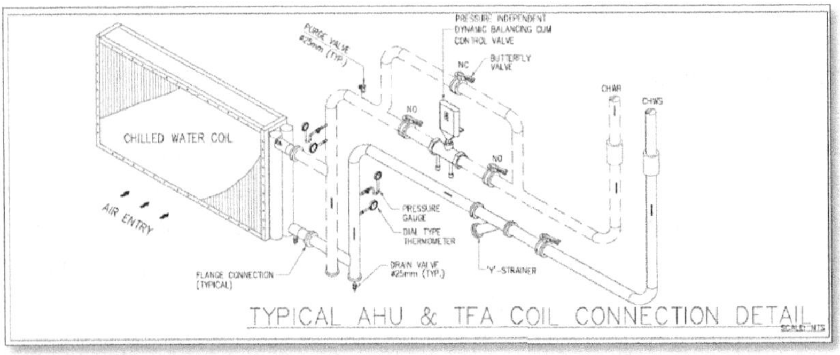

Figure 6.4 Typical AHU & TFA Coil Connection Details

- AHUS are typically dispatched in semi-knocked-down condition or in complete depending on the size of the unit. If it comes in semi-knocked-down condition, ensure the accessories are received in complete condition before starting to assemble. Ensure material is received as per the Packing list. If there is any deviation, inform OEM.

- Position the AHU on the foundation using spring mounts or inertia bases, or as per the contract documents and Method statements.

- Ensure alignment of fan, motor, and casing, level with the spirit level.

- Connect supply/return ducts with flexible connectors.

- Connect chilled water piping with vibration eliminators, strainers, and valves.

- Ensure access doors are oriented correctly for maintenance.

Fan Coil Units (FCUs)

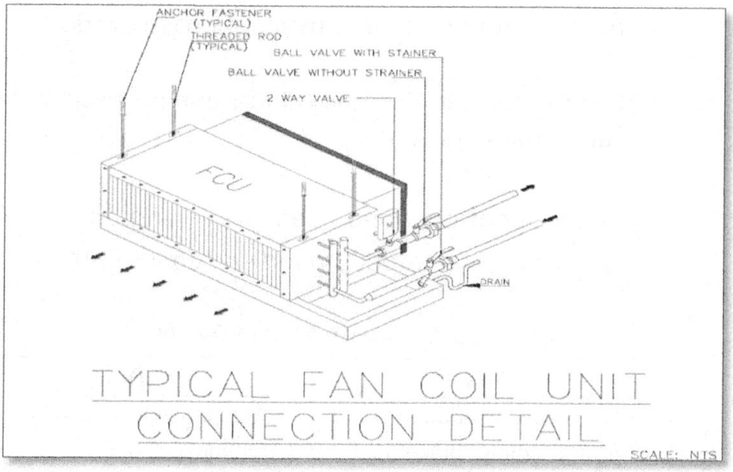

Figure 6.5 Typical Fan Coil Unit Connection Details

FCUs need to be ordered on the connection side, left or right. From the drawing, identify the chilled water pipe connection and place order accordingly. Do not rotate the coil at the site.

Verify suspended slab/wall supports with anchor bolts or brackets.

- Install vibration isolators (hangers or pads).
- Provide drain piping with proper slope and traps to avoid backflow.
- Connect refrigerant or chilled water piping as per drawings, with thermal insulation.

- Ensure electrical connections are as per approved submittals.

Treated Fresh Air Units (TFAUs)

- Place on foundations with vibration isolators.
- Connect fresh air intake duct, ensure proper louvres and filters are installed.
- Verify controls and sensors (temperature, CO_2, pressure) are properly wired and calibrated.

Table 6.5 Do's & Don'ts

✅ Dos	❌ Don'ts
• Always check foundation levels and isolator placement before equipment arrival. • Verify motor and fan alignment with gauges before commissioning. • Ensure strainers are flushed and cleaned before coil connections. • Provide a trap in drain piping with proper depth to avoid odour backflow. • Conduct pre-commissioning inspection with the consultant before energising.	• Don't install AHUs in unfinished, dusty environments without protective covers. • Don't connect ducts without flexible connectors — vibration will transmit through the system. • Don't leave inadequate space for maintenance — future serviceability will be compromised. • Don't skip leak testing of drain piping before handing over. • Don't bypass vibration isolators or gasket joints.

Golden Rule: *"Install today as if you are the one maintaining it tomorrow — service access is as important as system performance."*

6.3.5.5 Cooling Tower Installation and Condenser Water Piping:

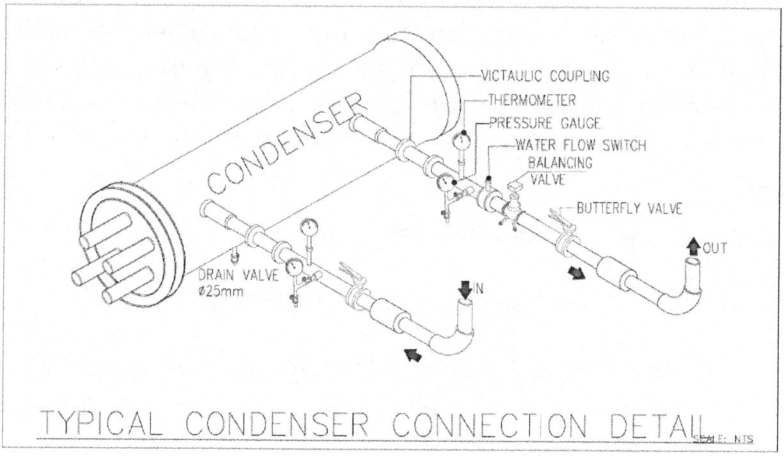

Figure 6.6 Typical Condenser Connection Details

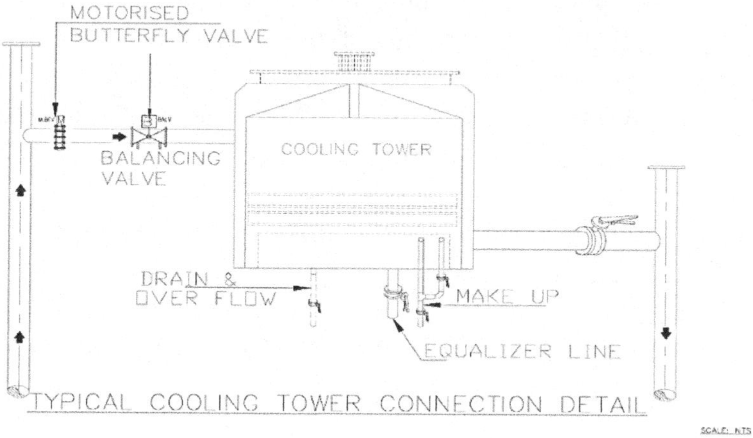

Figure 6.7 Typical Cooling Tower Connection Details

A cooling tower poorly installed is a lifetime of inefficiency — water wasted, energy lost, and performance compromised."

1. Introduction

Cooling towers are the heart of the heat rejection system in chilled water plants. Any error in their installation impacts chiller efficiency, energy consumption, and system reliability. Condenser water piping is equally critical — improper slopes, inadequate supports, or wrong valve selection can cause air entrapment, cavitation, or reduced system performance.

2. Pre-Installation Readiness

Before installation begins, verify the following:

- **Civil Works**: Tower foundation completed, RCC base or steel grillage checked for level and load-bearing capacity.
- **Access**: Crane/rigging access for lifting cooling tower sections and fan assemblies.
- **Vibration Isolation**: Anti-vibration pads/springs provided 3under the base.
- **Utilities**: Electrical cable trays, water supply, and drain points ready.
- **Water Treatment**: Provision for **Chemical dosing system**, bypass line, and sampling points planned.

3. Installation Procedure

3.1 Cooling Tower Assembly

- Unpack and inspect all tower parts (panels, fans, motors, drift eliminators, fill media).
- Place tower base/frame on foundation with vibration pads.
- Assemble casing panels, fill, eliminators, and fan stacks as per the manufacturer's manual.
- Install motor, gearbox, and fan with proper alignment and balancing.
- Connect the **make-up water line, overflow, and drain**.
- Provide access ladders, handrails, and platforms as per safety standards.
- Ensure isolator switch is at an accessible location

3.2 Condenser Water Piping Erection

- Route supply and return lines as per approved shop drawings.
- Provide eccentric reducers on the suction side of pumps (flat side up).
- Install strainers, balancing valves, temperature gauges, and pressure gauges at key points.
- Provide a balancing line if multiple towers are connected

- Ensure proper slope for air release and drain points at low levels.
- Provide pipe supports at regular intervals as per IS/ASHRAE standards.
- Apply thermal insulation (typically closed-cell elastomeric/nitrile rubber) to minimize heat gain and prevent condensation.
- Pressure test piping for any leaks

Table 6.6 Do's & Don'ts

✅ **Dos**	❌ **Don'ts**
• Check foundations and anchoring bolts before lifting tower. • Align motor and fan shafts precisely to avoid vibration. • Flush condenser water piping thoroughly before connecting to tower/chiller. • Provide air vents at high points and drain valves at low points. • Ensure all welds are inspected, hydro-tested, and approved. • Follow manufacturer's torque specs for fan blades and motor connections.	• Don't operate tower without proper water treatment — scale/corrosion will ruin efficiency. • Don't insulate piping before hydro-test — leaks will go undetected. • Don't neglect drift eliminators or louvers — they reduce water loss and noise. • Don't bypass balancing valves — uneven water flow reduces chiller efficiency. • Don't store pipes in open sun without protection — coatings/linings may degrade.

Golden Rule - *"Cooling tower and condenser piping are the lungs of the chilled water system — keep them clean, balanced, and protected to breathe efficiency into the plant."*

6.3.5.6 Plant Room Installation in HVAC Projects

1. Introduction

The **plant room** is the heart of any HVAC system, housing major equipment such as **chillers, pumps, cooling towers, heat exchangers, and headers**. Its correct layout and installation directly impact **system efficiency, ease of maintenance, and safety**. Since requirements differ, we will cover **air-cooled systems** and **water-cooled systems** separately.

2. Air-Cooled Chiller Plant Rooms

Air-cooled chillers eliminate the need for cooling towers and condenser water piping, making them suitable for **space-constrained projects** (like offices, malls, or smaller hospitals).

Key Considerations:

- **Location**: Install at the terrace or open-to-air areas for adequate heat rejection.
- **Clearances**: Maintain sufficient spacing between units for airflow and service access.
- **Vibration Isolation**: Use inertia bases and spring isolators under chillers and pumps.
- **Noise Control**: Acoustic barriers may be required in residential or hotel projects.

- **Electrical Panels**: Keep MCC/Control panels within the same plant area, with cable trays routed safely.

Figure 6.8 Air cooled Chiller

Installation Steps (Air-Cooled):

1. Prepare RCC foundation with anti-vibration pads.
2. Position chiller units with crane/lifting tools as per GA drawings.
3. Connect chilled water supply/return headers to pumps.
4. Install strainers, valves, and flexible connections.
5. Carry out pressure testing, flushing, and insulation of chilled water piping.
6. Integrate with control system (BMS/IBMS).

3. Water-Cooled Chiller Plant Rooms

Water-cooled chillers, more common in **large commercial complexes, hospitals, and industrial facilities**, involve additional subsystems like **condenser water pumps, cooling towers, make-up water tank**

Figure 6.9 Chilled water Pumps

4. Key Considerations:

- **Space Planning**: Separate areas for chillers, pumps, cooling towers, and electrical panels.
- **Hydraulic Balancing**: Ensure proper pump head calculations, balancing valves, and test points.
- **Pipe Routing**: Chilled water and condenser water piping should be segregated with colour coding.

- **Cooling Towers**: Located on a terrace or open yard, with proper drift eliminators and a water treatment plant (WTP).

- **Service Access**: Provide catwalks, lifting beams, and maintenance clearances.

5. **Installation Steps (Water-Cooled):**

1. Construct strong RCC foundations with anti-vibration bases.
2. Position chillers, pumps, and accessories as per plant layout drawings.
3. Install **chilled water (CHW)** and **condenser water (CW)** headers with expansion joints.
4. Connect cooling tower basin piping, makeup water lines, and bypass valves.
5. Carry out hydro-testing and chemical cleaning (flushing).
6. Provide thermal insulation for CHW and refrigerant piping.
7. Ensure all the controls are working properly
8. Integrate with BMS—include sensors for ΔT, flow, pump status, tower fan speed, etc.

Table 6.7 Do's & Don'ts

✅ Dos	❌ Don'ts
• Ensure GA (general arrangement) drawings are approved before starting installation. • Provide maintenance access (clear space for tube pulling, pump dismantling, etc.). • Use calibrated instruments for levelling, alignment, and flow testing. • Follow vibration isolation and acoustic control practices. • Verify equipment nameplates with approved submittals before installation.	• Don't overcrowd plant rooms — it reduces efficiency and makes maintenance impossible. • Don't install electrical panels in damp or leaking areas. • Don't bypass strainers, NRVs, or balancing valves. • Don't allow civil modifications without rechecking foundation load capacity. •

6.3.5.7 VRF Installation

Figure 6.10 VRF Installation

1. Introduction

While central chilled water systems dominate large-scale HVAC projects, **VRF (Variable Refrigerant Flow)** and **Packaged/Split AC units** are widely used in medium and small-scale buildings, or as supplementary systems. These installations offer **flexibility, energy efficiency, and faster deployment**. However, improper installation often leads to refrigerant leakage, poor cooling distribution, and shortened equipment life. In the last decade VRF installations have increased the market share primarily due to the flexibility they offer compared to chiller plants. With DX Coils of AHU getting connected to the VRF Outdoor units, these units are used even in hospitals and other large commercial air conditioning systems.

2. VRF Installation

Key Characteristics of VRF Systems:

- Uses refrigerant piping network (copper pipes) with branch selector boxes.
- Multiple indoor units connected to a single outdoor condensing unit.
- Requires precise refrigerant charge calculations and leak-proof piping.
- BMS/IBMS integration for advanced monitoring and energy optimisation.

Installation Steps:

1. **Outdoor Unit Placement**:

- Install on a terrace or external yard with adequate airflow and structural strength.
- Provide anti-vibration mounts and rain/sun protection.

2. **Copper Piping**:

- Route refrigerant pipes with approved insulation.
- Maintain bending radius, avoid excessive joints.
- Use only the approved refnet joints
- Perform nitrogen purging during brazing to prevent oxidation.

3. **Leak Testing & Pressure Testing**:
   - Carry out with nitrogen at the required pressures.
   - Hold test for a minimum 24 hours and check for drops.
4. **Vacuuming & Charging**:
   - Deep vacuum evacuation with micron gauge.
   - Refrigerant charging as per manufacturer guidelines.
5. **Indoor Unit Installation**:
   - Ceiling cassette/ducted/wall-mounted as per design.
   - AHU connection with the manufacturer's connector kit
   - Proper levelling, drainage connection, vibration isolation.
6. **Control Wiring**:
   - Install communication cables, integrate with BMS if applicable.

Best Practice: Always prepare as-built refrigerant piping drawings to avoid confusion during servicing.

3. Packaged / Split AC Units

Key Characteristics:

- **Split ACs**: Outdoor condensing unit + indoor evaporator unit. Common for small offices, cabins, and server rooms.

- **Packaged Units**: Larger capacity, pre-assembled with air handling components; used for medium halls, classrooms, retail areas.

Installation Steps:

1. **Outdoor Unit**:

   - Install on ground/terrace with vibration isolators.
   - Ensure free airflow and adequate service access.

2. **Indoor Unit**:

   - Proper location selection for uniform air distribution.
   - Fix securely with vibration isolation and slope drain piping.
   - Ensure adequate fresh air flow as per the design requirements
   - Proper space for maintenance is available.

3. **Copper Piping & Insulation**:

   - Ensure proper suction and liquid line installation.
   - Use appropriate insulation for refrigerant piping, protect against UV exposure.

4. **Drain Line**:

   - Test for water leakage, provide traps where required.
   - Give a proper slope for drain water flow

5. **Electrical & Controls**:
    - Ensure proper cable sizing, earthing, and safe routing.
    - Verify interlocking with thermostat/control panel.
6. **Testing**:
    - Run tests for cooling performance, noise level, drainage, and controls.

Table 6.8 Do's & Don'ts

✅ Dos	❌ Don'ts
• Follow manufacturer installation manuals strictly. • Pressure test and vacuum refrigerant lines before commissioning. • Ensure proper insulation of refrigerant pipes and drain lines. • Provide dedicated electrical circuits with MCBs/isolators. • Maintain clearances for service and airflow.	• Don't overcharge refrigerant — it reduces efficiency and damages compressors. • Don't route copper pipes longer than the manufacturer's recommended limits. • Don't leave drain lines without slope or traps. • Don't allow outdoor units in enclosed spaces with recirculated hot air. • Don't energise units without completing leak/pressure tests.

6.3.5.8 Electrical Panels & Controls

HVAC without controls is like a car without a steering wheel — it may run, but not in the right direction."

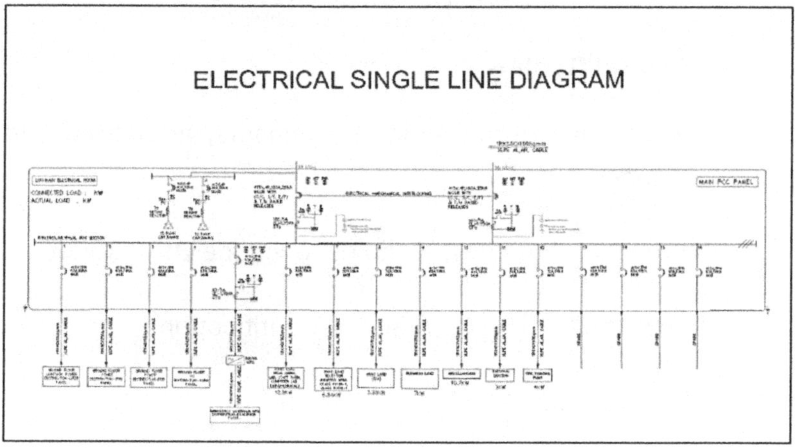

Figure 6.11 Electrical Single Line Diagram

1. Introduction

Electrical and control systems are the nervous system of any HVAC installation. While piping, ducts, and equipment form the body, it is controls that enable functionality, efficiency, and safety. Integration with Building Management Systems (BMS) or Intelligent Building Management Systems (IBMS) is critical for monitoring, automation, and long-term performance.

Poor execution here often leads to false alarms, control overrides, high energy wastage, and compromised safety.

2. Key Activities in Electrical & Controls Integration

1. **Power Cabling**

- Dedicated circuits for chillers, AHUs, pumps, cooling towers, and VRFs.

- Cable trays and routing coordinated with HVAC ducts and piping to avoid clashes.
- Earthing and lightning protection for sensitive equipment.

2. **Control Cabling**

- Low-voltage control wiring for sensors, actuators, and interlocks.
- Segregation from power cables to reduce interference.
- Proper tagging and ferruling for identification.

3. **Panels & Starters**

- Installation of MCC (Motor Control Centre) panels, VFD panels, and local starter panels.
- Verification of wiring as per single-line diagrams (SLDs).
- Ensuring accessibility for operation and maintenance.

4. **BMS/IBMS Integration**

- Interfacing of equipment (AHUs, chillers, VRFs, pumps) with the central control system.
- Uploading and testing of logic sequences (start/stop, safety interlocks, failover).
- Calibration of sensors (temperature, humidity, flow, pressure).
-

5. **Safety Interlocks**

- Fire alarm integration → auto-shutoff of AHUs, activation of dampers.
- Flow switches, pressure switches, and safety cutouts tested.
- Emergency stop pushes buttons validated.

Table 6.9 Do's & Don'ts

✅ Dos	❌ Don'ts
• Coordinate early with electrical/fire teams to freeze cable tray routes and interlock points. • Label all power and control cables clearly for easy identification. • Test sensors and actuators individually before linking them to the BMS. • Maintain separation of power and control cabling to avoid signal interference. Document all logic sequences, interlocks, and wiring diagrams.	• Don't bypass safety interlocks during testing —testing-this creates long-term risk. • Don't route cables haphazardly; always use trays or conduits with proper bends. • Don't ignore calibration; uncalibrated sensors lead to wrong readings and inefficient system control. • Don't hand over panels without training operators on alarm logic and reset procedures. Don't connect any cables loosely without suitable approved cable connectors

6.3.5.9 IBMS and IoT Integration

IoT and IBMS Integration with HVAC Systems "A building without intelligent controls is just a shell. IoT and IBMS make it responsive, efficient, and future-ready."

Apply IoT-enabled analytics and KPIs for continuous optimisation.

1. **Introduction The evolution of HVAC is no longer just mechanical—it is digital.** IoT (Internet of Things) and IBMS (Integrated Building Management Systems) are

transforming how buildings are operated, monitored, and optimised.

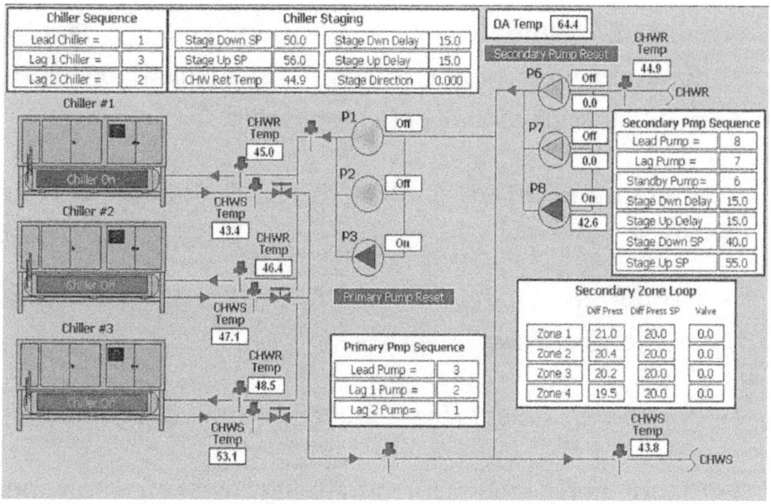

Figure 6.12 Typical Chiller Plant Monitoring System

Traditional systems focused on start/stop, safety interlocks, and manual monitoring, whereas today's IoT-enabled platforms provide real-time visibility, predictive analytics, and remote diagnostics. This reduces energy wastage, improves uptime, and ensures comfort without manual intervention.

1. Role of IoT for IBMS

- **Device Interoperability**: LonWorks enables seamless communication between diverse devices from different manufacturers, crucial for smart buildings.
- **Scalability**: Ideal for large-scale deployments across campuses or high-rise buildings due to its distributed control logic.

- **Real-Time Control**: Supports fast, reliable data exchange for real-time monitoring and automation of building systems.

- **LonTalk Protocol**: A media-independent, layered communication protocol designed specifically for control systems—not general data processing

2. **Key Elements of IoT & IBMS in HVAC Sensors & Field Devices**

- **Temperature**, humidity, CO_2, VOC (air quality), pressure, and flow **sensors**.

- Occupancy sensors for demand-controlled ventilation.

- Smart energy meters for monitoring power consumption.

3. **Controllers & Gateways**

- Local DDC (Direct Digital Controllers) for AHUs, chillers, and VRFs.

- Gateways for protocol conversion (BACnet, Modbus, KNX, MQTT).

- Edge computing devices for local analytics before sending to cloud.

Communication Protocols

- BACnet/IP, Modbus TCP, Local Operated Networks (LonWorks), KNX — for system-level communication.

- MQTT, REST APIs — for cloud/IoT integration.

4. **IBMS Platforms**

- Centralised dashboards integrating HVAC, lighting, fire, security, and lifts.
- Trend logs, alarm monitoring, and scheduled operation.
- KPI-based performance monitoring (e.g., kW/ton, COP, ΔT).

5. **IoT Cloud Integration**

- Data pushed to cloud platforms (Azure, AWS, Google Cloud) for storage and analytics.
- AI-based optimisation for chiller sequencing, AHU scheduling, and predictive maintenance.
- Mobile apps for real-time alerts and control.

6. **Benefits of IoT & IBMS Integration Energy**

- **Optimisation:** Auto-adjustment of AHUs and chillers based on occupancy and load.
- **Predictive Maintenance**: Early warnings for bearing failure, filter clogging, or pump vibration.
- **Remote Accessibility:** Facility managers can monitor and control systems anytime, anywhere.
- **Compliance & Reporting:** Automated generation of reports for audits, green building ratings, and energy benchmarks.

- **Occupant Comfort:** Better indoor air quality (IAQ) and personalised temperature control.

Table 6.10 Do's & Don't IoT/IBMS Integration

✅ Dos	❌ Don'ts
• Standardise protocols (BACnet, Modbus) across all vendors to avoid silos. • Ensure redundancy in servers and gateways to avoid single points of failure. • Validate sensor calibration before integration. • Train operators not just in panel handling but also in dashboard interpretation. • Archive trend data for at least 12 months for benchmarking.	• Don't overload the IBMS with unnecessary points — focus on critical KPIs. • Don't allow bypass of BMS alarms without proper authorisation. • Don't mix low-voltage (signal) cabling with high-voltage power cabling. • Don't rely on OEM dashboards alone — integrate into the central IBMS for a single view. • Don't ignore cybersecurity — unsecured IoT devices can be an entry point for hackers.

7. Case Example – Smart Campus Project A technology park in Hyderabad implemented an IoT-enabled IBMS integrating 10 chillers, 250 AHUs, and 2000 VRF indoor units.

Challenge: High energy bills, inconsistent cooling, and poor visibility across multiple buildings.

Solution:

- IoT sensors installed across AHUs and plant rooms.

- Cloud-based analytics for chiller sequencing and AHU scheduling.
- Dashboards tracking KPIs like kW/ton, system COP, and BMS alarms.

Result:

- 18% reduction in energy consumption.
- Faster fault detection and resolution (alarms within 5 minutes of the fault).
- Enhanced tenant comfort with demand-based cooling.

Lesson Learned: IoT/IBMS integration is not just an add-on; it is a performance enabler.

6.3.6: Measurements & Joint Records

Why it matters

Accurate measurements and joint records are the foundation of transparency in HVAC projects. Payments, variations, and claims are all tied to physical quantities installed at the site. Disputes often arise when joint measurements are not taken or poorly documented.

Key Practices

- **Joint Measurement Books (JMBs):** Maintained with signatures of contractor, consultant/PMC, and client.
- **Frequency:** At every milestone (ducting, piping, insulation, cabling).

- **Documentation:**

- Actual measurements

- BOQ reference item.

- Date, location, drawings referenced.

- Signatures of all parties.

> **Golden Rule:** *Mark the Actual measurements on the drawings.*

Common Pitfalls

- **Measurements delayed** → payments delayed → cash flow stress.

- Site works completed but no records → disputes later.

- Multiple versions of measurement formats → confusion.

6.3.7: Pre-Commissioning, Commissioning, Testing & Balancing

1. Pre-Commissioning

Activities that prepare the system for safe and effective startup:

- Flushing and chemical cleaning of chilled water piping.

- Pressure/leak testing of ducts, pipes, and vessels.

- **Electrical checks:** continuity, insulation resistance, grounding.
- Verification of safety controls (flow switches, pressure switches).
- Calibration of sensors and instruments.

2. Commissioning

Systematic process to start equipment and ensure it meets design intent:

- Start-up of chillers, AHUs, VRFs, cooling towers, and pumps.
- Verification of sequences: auto changeover, safety cutouts, alarms.
- Functional Performance Tests (FPTs).

3. Testing & Balancing (TAB)

- Air side: Airflow at diffusers, static pressure, fan curves.
- Water side: Pump head, ΔT across chillers/coil, flow balancing.
- System Integration: Confirm design parameters under full/part loads.

> **Golden Rule:** Never hand over without complete TAB reports — comfort and efficiency depend on it.

6.3.8: Integration with Commissioning

Why integration matters:

Commissioning is not only about equipment performance; it's about system-level integration. AHUs, chillers, pumps, VRFs, and controls must work together as a single ecosystem.

Key Elements:

- **IBMS/BMS Integration:** All equipment linked to central monitoring with points mapped and alarms tested.
- **Sequence of Operation Validation:** Confirm programmed logic matches design (e.g., pump lead–lag, chiller sequencing, AHU demand control).
- **Safety Interlocks**: Fire alarm shutdown, sump level interlocks, power failure restarts.
- **Data Logging & Trend Monitoring:** Minimum 7–14 days of data collection (kW/ton, ΔT, pump energy/m^3, BMS alarms).
- **Handover Training:** Operators trained on IBMS dashboards, trending, and alarm acknowledgement.

Golden Rule: *Commissioning is successful only when subsystems integrate seamlessly into the building's operations.*

Table 6.11 Do's & Don'ts

✅ Dos	❌ Don'ts
• Integrate commissioning into the project schedule early — treat it as a process, not an event.	• Don't push commissioning to the "last week" — rushed testing leads to incomplete validation.
• Ensure all pre-commissioning activities (flushing, leak tests, insulation, calibration) are documented and signed off.	• Don't energize equipment without completing pre-commissioning checks — it risks damage and warranty voiding.
• Conduct commissioning meetings with HVAC, electrical, controls, plumbing, and fire teams together.	• Don't let each trade test in isolation — lack of coordination leads to system-level failures.
• Validate system logic in BMS/IBMS (lead–lag, auto changeover, interlocks) with all stakeholders present.	• Don't assume programmed sequences work just because they're "in the software." Test them live.
• Use calibrated instruments for all TAB activities and archive results in commissioning reports.	• Don't rely on approximate/uncalibrated tools — wrong readings compromise performance.
• Perform seasonal commissioning tests (summer, monsoon, winter) and record results.	• Don't assume part-load or mild-season testing proves system performance.
• Train the O&M team with hands-on sessions using the actual IBMS dashboard, alarms, and instruments.	• Don't assume operators will "figure it out later" — lack of training leads to bypassing controls.
• Archive data logs (7–14 days minimum) for KPIs like kW/ton, ΔT, system COP, and BMS alarms.	• Don't skip trend logging — without data, disputes and inefficiencies cannot be resolved.
• Prepare and hand over complete commissioning documentation (FPT forms, TAB reports, system manuals).	• Don't close commissioning with incomplete paperwork — missing records cause disputes at handover.

"Commissioning is successful only when subsystems integrate seamlessly into one high-performing building system — with data, training, and documentation to prove it."

6.3.9 Commissioning Integration – At a Glance

Purpose

Ensure subsystems (air, water, controls, IBMS) integrate into one efficient, reliable building system that performs as per design intent.

Key Steps

1. **Pre-Commissioning** → Flushing, leak testing, insulation, calibration, readiness checklists.
2. **System Testing** → Equipment startup, subsystem functional tests, safety interlocks.
3. **Testing, Adjusting & Balancing (TAB)** → Air and water flows matched to design.
4. **IBMS/IoT Integration** → Lead–lag, auto-changeover, alarms, trend logging.
5. **Seasonal Testing** → Validate under peak and part-load conditions.
6. **Training & Handover** → Hands-on O&M training, calibrated tools, manuals.
7. **Data Logging & Reporting** → Archive performance logs for 7–14 days (KPIs).

KPIs to Track

- kW/ton of cooling
- ΔT across chiller/coil
- System COP (overall efficiency)
- Pump energy (kW per m^3/h)
- Number of BMS alarms (active/repeated)
- Hours of bypassed control/disabled VFDs
- Indoor air parameters (temperature, RH, IAQ)

Golden Rule: *"If it isn't tested, logged, and trained, it isn't truly commissioned."*

6.3.10 Checklists

- **Execution Readiness Checklist** (civil, materials, manpower).
- **HSE Checklist** (PPE, permits, training).
- **Commissioning Checklist** (pre-startup, TAB, handover docs).

- **Interface Management in HVAC Project Execution**

One of the biggest execution challenges in HVAC projects is **coordination with other trades** — civil, electrical, plumbing, interiors, and fire systems. Poor interface management is one of the top reasons for clashes, delays, and rework.

Typical Interface Issues:

- Ducts clashing with electrical cable trays or fire sprinklers.
- Wall openings not ready when duct/piping teams arrive.
- False ceilings closed before AHUs or diffusers are installed.
- Chiller foundations not ready for equipment delivery.
- Electrical power points delayed, preventing commissioning.

1. Best Practices for Interface Management:

✓ Hold weekly coordination meetings with all trades (civil, electrical, plumbing, interiors, fire).

✓ Use BIM models (if available) or shop drawing overlays for clash detection.

✓ Freeze shaft and ceiling cutouts well in advance.

✓ Define responsibility for scaffolding, supports, and wall openings in the contract or MoM.

✓ Maintain an **Interface Tracker** (who is responsible, status, due date, remarks).

✓ Escalate unresolved clashes early to the project manager/PMC.

Golden Rule: *"Interface gaps cause 80% of site delays. Manage interfaces, and you manage execution."*

2. Billing and Payments for Better Cash Flow

Even with flawless execution, HVAC projects can collapse financially if billing and collections are poorly managed. Delayed cash inflow cripples' procurement, vendor payments, and site progress.

Key Challenges:

- Clients are delaying certification of bills due to missing measurements or incomplete documentation.
- Excess quantities executed without amendments — contractors fund the gap until approval.
- Different billing cycles for vendors vs clients, creating negative cash flow.
- ***Retention money locked until project closeout.***

3. Best Practices for Billing & Cash Flow:

- ✓ Prepare monthly running bills with complete measurement sheets, signed jointly with the consultant/PMC.
- ✓ Get deviation **statements approved** immediately after take-off quantities are calculated.
- ✓ *Never send material beyond BOQ without a written amendment or client note.*
- ✓ *Align vendor payment cycles with client billing cycles wherever possible.*
- ✓ Track **Accounts Receivable (AR) vs Accounts Payable (AP)** weekly to prevent surprises.

✓ Maintain a **Cash Flow S-Curve** to monitor planned inflow vs actual inflow.

> **Golden Rule:** *"Execution without billing discipline is charity. Bill on time, collect on time, and fund your project sustainably."*

6.4 Conclusion – Project Execution

Project execution is the heartbeat of HVAC project management — where drawings, plans, and promises finally turn into ducts, pipes, equipment, and systems. Yet, execution is also where most projects stumble. Misaligned drawings, poor coordination, delayed material dispatches, skipped HSE protocols, or fragmented communication can erase the value of even the best planning.

A successful execution phase is not about simply "installing equipment." It is about:

- **Discipline in processes** — submittals, approvals, method statements, checklists.
- **Coordination across trades** — HVAC, civil, electrical, plumbing, automation, and IT.
- **Quality and safety in action** — ensuring installations meet standards and operators are protected.
- **Proactive monitoring** — measuring during execution, anticipating bottlenecks, and resolving before they escalate.

Execution is where the project manager's leadership is most visible: keeping the team aligned, ensuring material and manpower readiness, and maintaining a balance between speed, safety, and quality.

> **Golden Rule**: *"Execution is not just doing the work; it is delivering the design intent in reality, safely, on time, and at the right quality."*

Projects that respect this principle consistently achieve smoother commissioning, lower rework, and happier clients. Projects that cut corners in execution inevitably face delays, disputes, and reputational damage.

Execution is therefore not just a stage — it is the proving ground where professionalism, preparation, and performance meet.

Table 6.12 Project Execution – At a Glance (Summary Sheet)

Aspect	Key Points / Actions	HVAC Example
Purpose	• Convert planning and design into reality with safe, timely, and quality installation.	• From duct fabrication drawings → to ducts installed, tested, and ready for air balancing.
Inputs	• Approved drawings, technical submittals, method statements, WBS, schedule, BOQ.	• Latest IFC drawings, approved valve submittals, ducting BOQ.
Processes	1. Shop drawings & submittals 2. Site readiness & HSE compliance 3. Procurement & material delivery 4. Equipment/ system installation 5. Measurements & joint records 6. Pre-commissioning & commissioning 7. Integration with project monitoring.	• Duct installation → chilled water piping → AHU/FCU installation → VRF systems → plant room.
Outputs	• Installed & tested HVAC systems ready for commissioning. Documentation of inspections, approvals, NCRs, and quality records.	• Chiller foundation checklist signed off; AHU startup reports filed.

Best Practices	• Integrate execution with planning (update WBS, trackers). • Maintain HSE compliance daily. • Use checklists and inspection reports at every stage. • Train site engineers & operators during installation. • Escalate bottlenecks early (approvals, site readiness).	• Use joint measurement sheets to avoid disputes; daily safety toolbox talks.
Do's	• Follow approved method statements. • Ensure site readiness before material dispatch. • Conduct daily coordination meetings with other trades. • Use calibrated tools and proper supervision. • Document everything (MoM, approvals, NCRs).	• Safety clearance before scaffolding erection for duct installation.
Don'ts	• Don't treat execution as "just installation." • Don't skip documentation (shop drawings, MoM). • Don't dispatch excess materials without approval. • Don't bypass controls/BMS to "make it run." • Don't ignore operator training.	Commissioning was delayed because chillers arrived before the foundations were ready.

Golden Rules	• Execution = delivering design intent, safely and on time. • If it isn't documented, it doesn't exist. • Site readiness must be certified before material dispatch. • Commissioning begins with execution, not after it.	Contractor sends copper pipes early, rusts in the open yard → wastage.

Chapter 7: Testing and Commissioning HVAC Projects

"Commissioning turns promises on paper into measurable performance in the real world."

Learning Objectives

By the end of this chapter, the reader will be able to:

1. Understand the objectives of commissioning in HVAC projects.
2. Differentiate between pre-commissioning, commissioning, and post-commissioning activities.
3. Develop a Commissioning Plan starting from the planning stage.
4. Identify roles and responsibilities (PM, consultant, vendor, QA/QC, client).
5. Apply standard checklists, ITPs, and templates for commissioning HVAC equipment.
6. Recognise common pitfalls in commissioning and how to avoid them.
7. Appreciate the value of third-party commissioning and unbiased validation

7.1 Introduction

Commissioning is often misunderstood as the final act *of switching on the HVAC system and checking whether it produces cooling or heating as per design.*

Commissioning is a **comprehensive and systematic process** that begins during the planning stage and continues through execution, handover, and even into the first year of operation.

The **ASHRAE definition** describes **commissioning** as:

"A quality-focused process for enhancing the delivery of a project. The process focuses on verifying and documenting that the facility and all its systems and assemblies are planned, designed, installed, tested, operated, and maintained to meet the Owner's Project Requirements (OPR)."

In the Indian context, commissioning is often given less attention compared to design and installation. This leads to projects that may **technically "run"** but fail to **perform at their designed efficiency**. The result is increased energy consumption, higher maintenance costs, frequent client complaints, and, in some cases, expensive rework.

ISHRAE Standard concludes by saying, "*A HVAC system is generally designed for 1% Incidence. This means that the Design* assumes *that for 88 Hours in a year of 8760 Hours the load conditions peak simultaneously and at all other times the load conditions are milder. This effectively means that for 99% of the time, any fault in design, execution or selection of equipment can get covered up and does not become obvious.*"

Effective commissioning ensures that:

- The HVAC system delivers **thermal comfort, indoor air quality, and energy efficiency** as promised.
- Each piece of equipment is not only installed correctly but also **tested, balanced, and integrated** with the rest of the system. Each subsystem works at its designed efficiency. The complete system performance matches that of the design intent.
- **Owner's Project Requirements (OPR)** are met consistently under various operating conditions.
- Documentation, training, systems manual, and handover provide the client with confidence to operate and maintain the system sustainably.

ISHRAE also developed Standards for commissioning to educate on Commissioning processes right from pre-design stage to post handing over and operations.

ISHRAE Standard for commissioning emphasises, "*Commissioning is therefore not an isolated event but a **capable and systematic process of validating the design intent** and ensuring that the project delivers long-term value. When planned and executed properly, it transforms the project from a mere collection of equipment into a **well-integrated, high-performing system as per OPR**.*

7.2 Why Commissioning is Essential in HVAC Projects

Commissioning is not just about starting the HVAC system; it is about **ensuring that the design intent is transformed into a functioning reality**. Even with careful planning and design, system performance can be compromised due to a variety of factors:

- **Incorrect Design Assumptions:** During the design stage, assumptions are made about space use, occupancy, and load. If these assumptions prove wrong, the design can become ineffective. For example, a pump designed for a certain head and flow rate may operate inefficiently if site conditions alter system resistance.

- **Incorrect Materials Used:** At times, materials are procured without approval or substituted due to delays or non-availability. This deviation often reduces system efficiency.

- **Site Issues:** If the latest versions of drawings are not available at the site, installations may proceed using outdated versions. This results in rerouting of ducts, pipes, or cables, often leading to performance losses.

- **BMS Operation Bypassed:** Building Management Systems are sometimes bypassed during troubleshooting and never reinstated, leaving critical controls ineffective.

- **Maintenance Access Issues:** Inadequate space for maintenance, or missing access panels for concealed equipment, makes routine service impossible. Over time, this causes significant deterioration in performance.

Unfortunately, such issues are often discovered only when the system underperforms, leading to **rework, delays, and compromised quality**. Commissioning helps surface and resolve these problems early.

7.3 Scope of Commissioning

The scope of commissioning extends well beyond system startup. It begins as early as the pre-design stage and continues through design review, procurement, supply and storage, installation, and into operation.

It covers:

- Design reviews and verification
- Material procurement and storage checks
- Installation supervision and validation
- Pre-commissioning inspections (flushing, cleaning, leak testing)
- Startup process and functional testing
- Testing, adjusting, and balancing (TAB) of subsystems
- Review of system manuals and O&M documentation
- Training the owner's operations team (hands-on and classroom)
- Seasonal performance testing (summer/winter loads)

- Monitoring snags and closing them out
- Performance tracking before and after occupancy
- Updating lessons learned for future projects

7.3.1 Objectives of Commissioning

The commissioning process is designed to achieve the following:

- ☑ Convert the design intent into delivered reality.
- ☑ Ensure a structured, documented process throughout the project.
- ☑ Validate performance of subsystems and the overall system under different load and seasonal conditions.
- ☑ Ensure compliance with all relevant standards during design, procurement, and installation.
- ☑ Guarantee system performance for Indoor Environmental Quality (IEQ), occupant health, safety, water quality, and sustainability.
- ☑ Reduce long-term operation and maintenance costs.
- ☑ Deliver benefits to all stakeholders — owner, occupants, project team, and facility managers.

Golden Rule – Scope of Commissioning: *"Commissioning is not an isolated startup activity; it is a systematic, continuous process that ensures performance, efficiency, and client satisfaction."*

7.4 Commissioning Process Framework

Commissioning **runs in parallel with the HVAC** project framework — it is not an additional or isolated activity. The purpose of commissioning is to translate the **design intent into a delivered reality.**

The following Commissioning process framework is recommended:

Step 1: Develop Commissioning Scope / Define OPR (Owner's Project Requirements)

Step 2: Develop Commissioning Plan

Step 3: Prepare Inspection test plans (ITP), Functional Acceptance Test plans (FAT)

Step 4: Installation Validation

Step 5: Perform System Testing (equipment + system level)

Step 6: Performance Monitoring and Reporting

Step 7: Final handover with O&M manuals and training

According to ISHRAE, commissioning activities fall into three categories:

Table 7.1 ISHRAE categories

Category	Purpose
Initial Commissioning	Commissioning of subsystems and the overall system, functional testing, and seasonal performance validation.
Ongoing Commissioning	Fine-tuning during the warranty and post-occupancy phase; recording measurements and reporting performance.
Retro Commissioning	Re-commissioning of existing systems to validate design parameters and restore performance.

In this section, we focus only on the **Initial Commissioning Process**, as it lays the foundation for ongoing and future performance.

Step 1: Develop Commissioning Scope

Commissioning begins during the **pre-design phase**. The commissioning team, while independent of the project team, works in close collaboration to understand the **Owner's Project Requirements (OPR)**. These include both stated and unstated needs such as:

- Building type and application (hospital, school, commercial complex, etc.)
- Budgetary constraints and high-level schedule
- Site and environmental conditions (climate, water availability, utility costs)

- Applicable codes, standards, and green building rating requirements
- Energy efficiency and sustainability expectations (e.g., Net Zero goals)
- Indoor environmental quality, occupant health and safety considerations
- Reliability, maintainability, and integration with other building systems

Clearly defining the **scope and out-of-scope commissioning activities** prevents ambiguity and ensures alignment with the project's objectives

Step 2: Develop Commissioning Plan

Once the OPR and scope are finalised, the commissioning team develops a **Commissioning Plan** during the design and drawing stage, using the Design Basis Report as a foundation.

Key outcomes of the plan include:

1. Assurance that all subsystems and the entire system will be **planned, procured, inspected, installed, tested, and commissioned** before handover.
2. Confirmation that **key performance objectives** are measurable and achievable.
3. Identification of all **checklists, templates, and procedures** for project execution.

4. Development of **integration, system, and seasonal test plans**.

5. Establishing structured documentation for reporting and approvals.

Step 3: Prepare Inspection test plans (ITP), Functional Acceptance Test plans (FAT)

During the **detailed design and planning phase**, the commissioning team plays a proactive role:

- Validate that the design intent fully meets the OPR.
- Review equipment selection rationale, design assumptions, and performance specifications.
- Re-check key design calculations (e.g., pump head, AHU static pressure, duct sizing).
- Confirm serviceability and space for maintenance are included in the design.
- Verify adoption of applicable, affordable, and available technologies.
- Review technical submittals, method statements, and report deviations.
- Prepare detailed inspection plans for raw materials and **factory acceptance testing (FAT)**.
- Develop equipment installation validation checklists.

- Align the commissioning schedule with equipment deliveries and construction activities.

Step 4: Installation Validation: Construction Phase

The construction phase is where **commissioning gains maximum visibility**. The commissioning team ensures quality, safety, and compliance through:

- Verifying safety procedures, tools, and training at site.
- Inspecting materials delivered to site against approved submittals; recording nameplate details, factory tests, and conducting random inspections of ducts, supports, and pipes.
- Checking installation of high-side equipment (chillers, cooling towers, pumps) as per contract and method statements.
- Ensuring plant rooms and AHU rooms provide adequate service and maintenance space.
- Observing welding processes, insulation application, duct and pipe routing.
- Verifying installation of accessories and test points needed for balancing and performance measurement. (Many Indian sites miss these, resulting in rework or inaccurate testing later.)

Step 5: Perform System Testing

System testing is not limited to startup and performance verification — it also includes preparing system documentation, and ensuring all deficiencies (snags) are resolved.

Key activities in this stage include:

- **Unit and System-Level Testing**
  - Startup checks for chillers, AHUs, VRF systems, pumps, and cooling towers.
  - Functional Performance Tests (FPTs) for each subsystem.
  - Testing, Adjusting, and Balancing (TAB) of air and water systems.
  - System-level integration tests: control sequences, BMS interfacing, safety interlocks.
  - Seasonal performance testing under realistic operating conditions.
  - Document results, deficiencies, and corrective actions

Note: Please refer ISHRAE Commissioning Handbook for the detailed testing and commissioning activities

Step 6: Performance Monitoring and Reporting

Commissioning is not complete once the system is tested, balanced, and handed over. A major portion of the process continues during **day-to-day operation**, where system performance is validated against the **Basis of Design (BOD)** under real load conditions.

During operation, many factors can cause system performance to drift away from design intent:

- Natural wear and tear of equipment
- Fouling of cooling coils and heat exchangers
- Improper operation of actuators, dampers, and sensors
- Variable Frequency Drives (VFDs) being bypassed or disabled
- Uncalibrated measuring instruments
- Poor or irregular maintenance practices

6.1 Data Logging, KPIs & Trend Monitoring for HVAC Commissioning

Why Data Logging?

While commissioning validates performance at a point in time, continuous data logging over **7–14 days** ensures the system performs consistently under varying load and environmental conditions. Logged data should be analysed and archived in the Commissioning Report.

Table 7.2 Commissioning Report

KPI	Description	Unit / Formula	Purpose
Plant Efficiency (kW/ton)	Power consumed per ton of refrigeration delivered.	kW input / TR delivered	Confirms system energy efficiency.
Chilled Water ΔT	Difference between chilled water supply & return temperatures.	$\Delta T = T\_return - T\_supply$	Confirms design heat transfer effectiveness.
System COP (Coefficient of Performance)	Ratio of cooling output to power input.	COP = Q_cooling (kW) / Power input (kW)	Verifies overall efficiency.
Pump Energy per m^3	Pumping energy consumed per cubic meter of water delivered.	kWh / m^3	Confirms hydraulic efficiency.
BMS Alarms Count	Number of alarms generated during the monitoring period.	Count/day	Indicates system stability and reliability.
Bypassed Control Hours	Hours where BMS/controls were disabled or overridden.	Hours logged	Reveals improper operation practices.
Airflow Balancing Accuracy	The difference between design and measured airflow.	% deviation	Validates TAB results.
Humidity & Temperature Stability	Range of indoor conditions during occupancy.	°C and %RH trends	Confirms occupant comfort and IAQ.

6.2 Best Practice for Data Logging

- Use calibrated data loggers and trending software linked to BMS wherever possible.

- Logging duration: 7–14 continuous days (covering peak and partial loads).
- Sampling interval: 5–15 minutes for power, flow, and temperature.
- Archive logs in the Final Commissioning Report (digital + hard copy).

Share dashboards with the O&M team for capacity building.

> **Golden Rule:** *"A system may 'run' on the day of handover, but only logged data proves that it performs consistently, efficiently, and as per design intent."*

The commissioning team (or designated authority) should:

- Review **system logs and maintenance records regularly**.
- Analyse subsystem efficiency and overall system performance under daily loads.
- Verify if **preventive maintenance schedules** are being followed and are effective.
- Perform detailed calculations to confirm whether performance aligns with design intent or has drifted.
- Conduct **root cause analysis** of any deviations.
- Recommend and track corrective and preventive actions.

6.3 Outcomes of Performance Monitoring:

- Regular analysis of system performance and deviations.
- Identification of root causes for inefficiencies.
- Recommendations for corrective and preventive actions, along with monitoring of implementation.
- Performance reporting to stakeholders, ensuring transparency and continuous improvement.

Golden Rule – Performance Monitoring: *"A system that is not monitored will eventually underperform — and no amount of commissioning at handover can prevent that."*

Step 7: Training and Handover:

In the HVAC industry, it is unfortunately common to hand over systems to untrained or inexperienced operators. While these operators may claim prior experience with similar plants, every project has unique design intents, parameters, and control philosophies. Without proper training, the system may appear to be "running," but there is no assurance it is operating at the designed efficiency or within the intended energy consumption levels.

Too often, due to a lack of knowledge, operators bypass controls, disable sensors, or override Building Management Systems (BMS) simply to keep the plant running condition. This practice not only defeats the very purpose of installing sophisticated control systems but also results in higher energy consumption, increased wear and tear, and premature equipment failure.

Golden Rule: "A running plant is not always a performing plant — commissioning and operator training ensure the difference."

7.1 Training of Owner's O&M Team

- Conducted by the Commissioning Authority (CxA) in coordination with the design and installation teams.
- Initial training during construction on the OPR and Basis of Design, including system limitations.

CxA organises a training development workshop to finalise the scope and content of training sessions.`

- Training ensures the O&M team is capable of operating, optimising, and maintaining the system as per OPR.
- Training must be documented, and the competency achieved by participants should be verified against OPR.

7.2 Preparation of Systems Manual

- A final commissioning report is prepared summarising the entire commissioning process.
- Systems manuals must be complete before training is delivered.
- Content includes:
  - Design Basis Report (DBR) and approved specifications
  - As-built drawings
  - Materials and equipment list with serial numbers

- Factory test certificates and inspection results
- Installation records and photographs
- System operating procedures (including dos & don'ts, prerequisites)
- Training records and attendance
- Overall system performance reports
- Brochures and technical information of key equipment

Golden Rule: System Manual - *"A system manual that isn't used is as useless as a tool that's never calibrated.*

7.3 Snag Identification and Removal.

- Deficiencies identified during commissioning must be logged, assigned, and tracked until closure.
- Post-occupancy testing should be conducted to verify system performance once the space is operational.
- All snag closures and retests must be documented in the commissioning report.

7.5 Tools & Templates for Commissioning

Commissioning without structured tools often results in missed steps, inconsistent reporting, and disputes later. Simple, well-designed templates and calibrated instruments ensure that commissioning is both systematic and transparent.

7.5.1 Documentation Tools & Templates for Commissioning

- **Commissioning Plan Template**

  Outlines commissioning objectives, roles, responsibilities, milestones, and reporting structure. It ensures all stakeholders are aligned before execution begins.

- **Pre-Commissioning Checklists**

  Prepared for major equipment like chillers, AHUs, and VRF systems. These checklists verify readiness before startup — foundations, piping flushed, strainers cleaned, insulation completed, electricals and controls checked, and safety interlocks tested.

- **Functional Performance Test (FPT) Forms**

  Used to record the results of performance testing under different operating conditions. They validate that equipment and systems perform as per design. Each test result is documented, compared against expected values, and marked pass/fail.

- **Commissioning Tracker**

  A consolidated log to monitor the status of commissioning activities. Tracks what is completed, pending, or needs rework, along with responsible parties and due dates.

7.5.1.1. Commissioning Plan Template (Word/Excel)

- Outlines the scope, objectives, schedule, and responsibilities of commissioning activities.
- Sections: OPR, roles & responsibilities, commissioning team, schedule milestones, reporting format, escalation process.
- **Use:** Keeps commissioning structured and transparent from planning to handover.

7.5.1.2. Pre-Commissioning Checklist (Excel)

- Equipment readiness checks before startup.
- Example for **Chiller**:
  - Foundation completed & vibration pads installed.
  - Piping flushed, strainers cleaned.
  - Electrical connections verified.
  - Safety controls checked (high/low pressure cutouts, flow switches).
- **Use:** Ensures no critical steps are missed before energising systems.

7.5.1.3. Functional Performance Test (FPT) Form (Excel/Word)

- Records results of functional testing under various operating conditions.
- Example fields: Equipment ID, test condition, expected result, actual result, pass/fail, remarks.
- **Use:** Validates that systems perform to design intent (e.g., AHU airflow, pump flow/head, VRF unit performance).

7.5.1.4. Testing, Adjusting & Balancing (TAB) Report Format (Excel)

- Captures measured air and water flow rates vs design values.
- Includes balancing adjustments and final readings.
- **Use:** Confirms that airflow/waterflow is balanced across the system to ensure comfort and efficiency.

7.5.1.5. Commissioning Issues Log / Deficiency Tracker (Excel)

- Records snags or deficiencies observed during commissioning.
- Columns: Issue ID, description, subsystem, responsible party, target date, status, closure date.
- **Use:** Provides accountability and tracks closure of all issues before handover.

7.5.1.6. Training Attendance & O&M Manual Verification Form (Word)

- Tracks training sessions conducted for the client's O&M staff.
- Verifies O&M manuals, warranties, and as-built drawings have been handed over.
- **Use**: Ensures smooth transfer of knowledge to the operations team.

7.5.1.7. Seasonal Commissioning Checklist (Excel)

- Records performance verification during different seasons (summer/winter/monsoon).
- Example:
  - Summer – confirm chilled water system capacity.
  - Monsoon – verify humidity control and IAQ.
  - Winter – confirm heating performance if applicable.
- **Use:** Ensures the system performs across varying conditions, not just at handover.

Golden Rule – Commissioning Tools: *"If it isn't documented, it isn't commissioned."*

7.6 Measuring Tools & Instruments

No commissioning process is effective without the right measuring tools. Instruments ensure accuracy, repeatability, and transparency of test results.

Air Balancing Tools:

- **Anemometer (hot-wire / vane):** Measures air velocity and airflow at diffusers and ducts.
- **Capture Hood :** Measures airflow at grilles and diffusers directly.
- **Manometer / Differential Pressure Gauge:** Measures duct static pressure, filter pressure drop, and fan performance.
- **Smoke Generator / Smoke Tube:** Visualises air movement and leakage.

Water Balancing Tools:

- **Ultrasonic Flow Meter:** Non-invasive measurement of water flow in pipes.
- **Differential Pressure Gauge:** Measures pressure drops across balancing valves to calculate water flow.
- **Temperature Sensors / Probes:** Verifies ΔT across chillers, AHUs, and coils.
- **Portable Water Test Kits:** Checks water quality parameters like pH, hardness, and corrosion inhibitors.

General Testing & Electrical Tools:

- **Clamp Meter / Multimeters:** Measures electrical load, voltage, and motor current.

- **Data Loggers:** Records temperature, humidity, and energy use over extended periods.
- **Sound Level Meter:** Confirms compliance with Noise Criteria (NC) ratings.
- **Infrared Thermometer / Thermal Camera:** Identifies insulation gaps, hot spots, and abnormal heating in equipment.

All instruments should be regularly calibrated, and calibration records should be maintained as part of commissioning documentation.

> ***Golden Rule – Commissioning Tools:*** *"One calibrated set of tools and instruments used during commissioning must be handed over to the operating team. Just as important, the operators must be trained to use them correctly whenever system checks or adjustments are required."*

7.7 Dos and Don'ts of Commissioning

Table 7.3 Dos and Don'ts of Commissioning

✅ Dos	❌ Don'ts
Integrate commissioning into the project schedule from the planning stage — **treat it as a process, not an event.**	**Don't treat commissioning as just "switching on"** equipment on the last day.
Ensure all **technical submittals, test records, calibration logs, and O&M manuals are completed, signed, and archived**.	Don't proceed with commissioning if documentation is incomplete or missing.
Conduct commissioning meetings with HVAC, electrical, plumbing, and controls teams together. Use a **clear responsibility (RACI) matrix.**	Don't allow trades to work in silos — lack of coordination leads to integration failures.
Plan **for seasonal load tests** (summer, monsoon, winter) and include them in the commissioning schedule.	Don't **hand over the system** without verifying performance under all expected **seasonal/part-load conditions.**
Provide **structured training to the O&M** team, including tool handling, parameter ranges, and troubleshooting. Record attendance and competency.	Don't assume operators will "figure it out" — skipping training guarantees inefficiency and misuse of controls/BMS.

Golden Rule – Pitfalls: *"Most commissioning failures are not technical — they are due to lack of planning, documentation, and training."*

7.8 Conclusion

Commissioning is not the final checkbox in an HVAC project — it is the **thread that runs from design to operation**, ensuring that every stage delivers what was intended.

It validates design assumptions, uncovers site realities, ensures correct installation, and translates theory into performance. Without it, systems may appear operational but underperform, leading to inefficiencies, occupant discomfort, and frequent rework.

A **well-structured commissioning process** provides:

- Confidence to the owner that the investment is secure.
- A reliable and efficient system for the operations team.
- Lower operating costs and fewer surprises for the facility.
- Peace of mind for end-users who experience comfort and reliability.

> **Golden Rule**: *In essence, commissioning is the difference between a project that looks complete on paper and one that performs flawlessly in real life.*

7.9 Commissioning at a Glance

Scope

- Starts at pre-design, continues through design, procurement, installation, testing, and handover.
- Covers initial, ongoing, and retro commissioning.

- Extends into post-occupancy monitoring and seasonal performance validation.

Objectives

- Convert **design intent into delivered performance**.
- Ensure compliance with OPR, DBR, and standards.
- Optimize energy efficiency, comfort, and reliability.
- Reduce O&M costs and improve long-term sustainability.
- Equip owner's team with training and complete documentation.

Key Steps

1. Develop Commissioning Scope (OPR).
2. Prepare Commissioning Plan (DBR, checklists, test plans).
3. Validate design and prepare inspection/testing plans.
4. Oversee installation with inspections and validation.
5. Conduct system testing, balancing, training, and snag closure.
6. Monitor performance during operation, analyze logs, and ensure preventive actions.

Golden Rule - *"Commissioning is not a one-day test — it is a disciplined process that transforms drawings into performance and promises into reality."*

Chapter 8: Monitoring and Controlling

"What gets measured, gets managed. What gets ignored, gets delayed."

Learning Objectives

By the end of this chapter, the reader will be able to:

1. Understand the importance of monitoring and control in HVAC projects.
2. Apply methods to measure and manage schedule, cost, scope, and quality variances.
3. Use Earned Value Management (EVM) to evaluate project health.
4. Track procurement, risk, and issues systematically.
5. Establish effective communication and reporting mechanisms.
6. Integrate monitoring of commercials, compliance, and HSE performance.
7. Apply practical tools, templates, and dashboards to control projects in real time.

8.1 Introduction: Why Monitoring & Control Matters

"That which cannot be measured cannot be controlled or improved."

Monitoring is the process of collecting, recording, and reporting project performance information that is shared with different stakeholders as per their needs. Controlling, on the other hand, is the act of initiating corrective action based on this performance information.

For example:

- Recording the progress of ducting, piping in different zones is **monitoring**.
- If ducting is behind schedule, increasing the team size or adjusting shifts to recover time is **controlling.**

Studies indicate that only 25–**30% of projects** follow detailed monitoring against their schedules. Nearly 70**% of schedules are never updated once prepared,** and more than 50% are never baselined. Without monitoring and baseline control, project schedules become little more than paperwork.

Scope management adds another layer of complexity. Small-sounding changes, such as requesting an additional air outlet in a room, may appear trivial but can trigger chain reactions: re-routing ducts, adding material, revising BOQ, delaying installation, and impacting procurement timelines. Similarly, in HVAC projects, numerous dependencies make control critical:

- Consultant approvals for technical submittals.
- Vendor POs and factory acceptance tests.
- Material delivery to site (with customs delays for imports).
- Site readiness (plant rooms, shafts, wall openings).

- Availability of power and water for testing.

A well-monitored project identifies such bottlenecks early and resolves them before **they snowball into costly disruptions.**

Golden Rule: *"A beautifully prepared schedule without monitoring is just a wish list."*

8.2 Variance Measurement and Control

Project phases —**planning, execution, and monitoring/ control** — form a continuous loop. Planning does not end once the baseline schedule is frozen. During execution, several **variations from the plan occur** due to the inherent risks in any project. These require corrective actions, and the cycle of planning–monitoring–controlling continues until project completion. Unless the project manager has a smoothly **operating monitoring/control system**, it is extremely difficult to manage the project effectively.

In any HVAC project, schedule**, cost, and scope variations** form the primary basis for measurement and control. PMBOK recommends that other knowledge areas — such as **quality, risk, procurement, resources, communication, and integration management** — are also monitored to ensure overall project success.

Golden Rule: *First measure, then analyse, then control.*

8.2.1 Schedule Variance (SV)

How to measure:

- Update every activity in the **Gantt chart/WBS** as soon as it is completed.
- Record actual hours/days, actual date of completion, and resources used.
- Automate data capture where possible and train the PMO to update information at fixed intervals.
- Use SPI (Schedule Performance Index = EV / PV) for analysis.

Proactive Monitoring During Execution

Measuring progress while the activity is in progress is far more effective than waiting until delays occur. Proactive monitoring allows early identification of bottlenecks and ensures that tasks are completed on schedule.

Example: Late approval of duct fabrication drawings delays duct installation, which in turn postpones false ceiling works. Similarly, duct installation can get stalled due to missing wall openings or a lack of clarity on who is responsible for scaffolding. These factors may not always be captured explicitly in the project schedule, but site engineers and supervisors must anticipate them and plan ahead for smooth execution.

Contractors should maintain a comprehensive checklist of requirements—including even small items like bolts and nuts—that could otherwise become reasons for unnecessary stoppages and project delays.

Control Measures:

- Re-sequence non-dependent activities.
- Increase manpower or shifts.
- Escalate bottlenecks (e.g., site not ready, clashes with other trades).

Golden Rule: *Time lost early in the project is never fully recovered unless action is swift.*

8.2.2 Cost Variance (CV)

How to measure:

- **Material costs:** From invoices submitted along with deliveries.
- **High-side equipment:** Include advances paid for chillers, VRFs, etc., as per accounting standards.
- **Resource costs:** Arrived at by recording man-hours × average cost per resource. (Monitoring uses averages, accounting uses actuals).
- **Use Earned Value Management** (CPI, Cost Performance Index = EV / AC) for analysis.

Note: EVM is explained in detail later in this chapter

Example: Excess ordering, insulation → excess inventory → wastage and higher costs.

Control Measures:

Effective cost control in HVAC projects requires discipline in procurement, material movement, and site coordination. Variances often arise not from major errors but from small, unchecked actions that snowball into cost overruns. The following measures help keep costs aligned with the approved BOQ and budget:

Control Measures

- ✓ **Strict Procurement Monitoring:** Track purchase orders closely and link them to approved quantities in the BOQ.
- ✓ **Material Dispatch Control:** Never send materials to site in excess of BOQ quantities without a formal amendment or written client approval.
- ✓ **Site Readiness Check:** Dispatch sensitive and high-value materials (e.g., copper pipes, GI sheets, valves, insulation) only when the site is certified as ready for installation.
- ✓ **Regular BOQ Reconciliation:** Compare actual issued, installed, and measured quantities against BOQ on a regular basis (weekly or monthly).
- ✓ **Rework Cost Tracking:** Record rework separately and obtain client/consultant approval wherever possible, to avoid eating into project margins.

> **Golden Rule:** *"Every material movement must have two approvals — one from the BOQ and one from the site readiness checklist."*

8.2.3 Scope Variance

How to measure:

- Scope is measured against the BOQ.
- Materials are tracked as received, stored, installed, and commissioned.
- Actual scope is typically measured at installation stage, since only erected quantities are payable in most of the contracts. (Note: Please refer to the contract clause for measurement.)
- Joint measurements are recorded by PMC, consultant, and contractor for variable quantities (ducting, piping, insulation, cabling).
- Agreed measurement standards (SMACNA, IS, or client-specific) should be frozen before the project begins to avoid disputes.
- In some government contracts with lump **sum quantities**, contractors bear the risk of variation — a case of risk transfer.

Example - How Small Scope Changes Escalate into Variances

A client requested an **additional air outlet** after the drawings were already frozen.

- This required a **duct extension**, new **diffuser**, and **system rebalancing**.

- At the same time, the routing of **electrical cabling** was altered to a longer path, resulting in a **shortage of cables on site**.

- These small deviations, though seemingly minor, triggered material shortages, delays in site execution, and ultimately placed **commissioning on hold** until the issues were resolved.

Lesson Learned: Even minor client-driven changes after freeze points can cause disproportionate cost and schedule impacts. Every variation must be **logged, approved, and linked to BOQ amendments** before site execution.

Control Measures:

- ➢ Implement a Change Control Board (CCB) process.
- ➢ Record impact on time and cost before approving changes.
- ➢ Communicate scope implications to all stakeholders.

Golden Rule: *No "additional work beyond the Agreement", without an amendment / written note from the customer.*

8.2.4 Quality Variance

Measuring Quality Variance in HVAC Projects

Quality variance occurs when the installed product or process deviates from the approved standards, technical submittals, or quality plan. In HVAC projects, these variances can directly impact system efficiency, safety, and project timelines.

8.2.4.1 How to Measure Quality Variance

- **Technical Submittals as Benchmarks** → Define quality parameters and acceptable ranges.

- **Physical Inspections** → Use instruments like Vernier calipers, screw gauges, and pressure gauges to verify dimensions, thickness, and tolerances.

- **Quality Control Tests** → Perform checks as per the approved Quality Plan (e.g., leakage testing, pressure testing, insulation resistance tests).

- **Factory Acceptance Tests (FAT)** → Confirm equipment compliance before dispatch.

- **Commissioning Stage Tests** → Verify quality in real operating conditions via staged testing, balancing, and performance verification.

- **Process Quality Audits** → Examine adherence to approved processes (e.g., "Was material dispatch done per SOP?" → If not, identify reasons, record variance, and raise CAPA).

- **Non-Conformance Reports (NCRs)** → Capture deviations, track corrective/preventive actions, and close with approvals.

Example – Process Lapse Causing Quality Variance

A welding activity was carried out by an **untrained worker**, leading to a **joint failure** during testing.

- The weld was rejected.
- Rework consumed time, material, and delayed subsequent activities.

Lesson Learned: A single skill gap at the execution level can snowball into major rework and cost impacts.

8.2.4.2 Control Measures

- Maintain quality checklists and inspections at every stage.
- Provide regular training for site engineers, supervisors, and technicians.
- Empower QA/QC engineers with the authority to stop/reject non-compliant work.
- Use process audits to monitor how work is being done, not just the end result.
- Track **NCRs** with root cause analysis and CAPA closure.

> **Golden Rule – Quality Variance:** *"Don't just measure the final product — measure the process. Consistent process control prevents inconsistent outcomes."*

8.2.5 HSE Integration in Monitoring & Control

HSE (Health, Safety, and Environment) is not separate from project monitoring — it is an integral part. A project delivered on time, but with accidents or unsafe practices, cannot be considered a success. Monitoring HSE ensures that safety standards are embedded into daily execution, not treated as a compliance formality.

What to Monitor in HSE

- **Leading Indicators** (predictive)
  - Toolbox talks conducted.
  - Safety inductions for new workers.
  - PPE usage compliance % observed in audits.
  - Number of safety observations raised & closed.
- **Lagging Indicators** (after-the-fact)
  - Near-misses reported.
  - First-aid cases.
  - Lost Time Injury (LTI).
  - Reportable incidents.

Golden Rule: *"Unsafe progress is not progress — a project delayed for safety is still a project saved."*

8.2.6 Issue Log: Purpose and Use in Complex HVAC Projects

In large HVAC projects, dozens of small issues crop up daily — delayed approvals, missing site access, vendor delivery slippages, or clashes with other trades. If left undocumented, they are forgotten until they resurface as major risks or disputes.

An **Issue Log** is a central tool to capture, assign, and resolve such issues systematically.

- **Purpose of an Issue Log:**
  - Records all project issues in one place.
  - Assigns ownership (who is responsible for resolving it).
  - Tracks due dates and resolution status.
  - Provides quick visibility for escalation when issues remain open too long.
  - Reminds ownership among stakeholders
- **In Complex Environments:**

  In multi-agency HVAC projects (civil, electrical, interiors, IT, fire), the issue log prevents finger-pointing. Instead of arguments, each issue has:

  - A unique ID.
  - Description.
  - Assigned owner.
  - Target resolution date.
  - Closure remarks.

Golden Rule: *"What gets recorded gets resolved."*

Table 8.1 – Issue Log

Issue ID	Description	Owner	Target Date	Status	Escalation Needed	Closure Remarks
ISS-01	**AHU submittal pending consultant approval**	Design Engineer	10-Mar-25	Open	Yes → PMO	Bulk approval workshop planned.
ISS-02	**Scaffolding not available in Block A**	Site Engineer	05-Mar-25	Closed	No	Scaffold erected, work resumed.
ISS-03	**Copper pipe shortage – revised route**	Procurement Lead	12-Mar-25	Open	Yes → Client	Amendment request submitted.
ISS-04	**Delay in civil handover of shaft-2**	Civil PMC	15-Mar-25	Open	Yes → Steering Cmte	Awaiting clearance.

8.2.7 ☑Dos and ✗ Don'ts of Effective Project Monitoring

Table 8.2Dos and Don'ts of Effective Project Monitoring

☑ Do's	✗ Don'ts
Monitor other Key areas: Procurement, Communication, Resources, Integration, Commercials, Compliance, and HSE. **Update regularly** – Keep schedules, logs, and trackers current (weekly, or more often for fast-track jobs). **Focus on the critical path** – Track activities that directly impact project completion. **Train your team** – Ensure PMO staff and site engineers know how and why to update trackers accurately. **Visualise performance** – Use charts, dashboards, and trend lines to make insights actionable. **Escalate early** – Raise minor issues before they snowball into major delays.	**Don't assume:** Get your assumptions clarified . (e.g., Chiller dispatched ≠ Chiller delivered to site). **Don't let updates slide** – Outdated data gives false confidence and hides risks. **Don't monitor everything equally** – Treating all tasks as critical dilutes focus. **Don't assume staff know the tools** – Lack of training leads to wrong or inconsistent data. **Don't drown in spreadsheets** – Long lists without visualisation hide problems instead of solving them. **Don't wait until a crisis** – Silent monitoring often ends in firefighting.

8.2.8 Dashboards: Online vs Offline

A **dashboard** is the project's "control room", similar to a car dashboard that tells the driver whether the journey is safe and on track.

- **Offline Dashboards (Excel-based):**
  - Best for small-to-medium HVAC projects.

- Easy to customise with formulas, conditional formatting, and simple Gantt charts.
- Can integrate BOQ, invoices, and procurement trackers.
- **Limitation:** Manual updates, prone to errors if not maintained diligently.

- **Online Dashboards (Cloud/ERP-based):**
    - Ideal for large and multi-location projects.
    - Tools like MS Project Online, Primavera P6, Zoho Projects, Monday.com, or ERP-integrated dashboards.
    - Advantages: Auto-updates, role-based access, mobile app inputs, and real-time collaboration.
    - **Limitation:** Subscription costs, training requirements.

Golden Rule: *"A dashboard is only as good as the accuracy of the data feeding it."*

8.2.9 Conclusion– Monitoring & Controlling

Monitoring and controlling are not about micromanaging people or pointing out failures — they are about **creating visibility, accountability**, and timely opportunities for correction. In HVAC projects, where even a single delay in ducting, piping, AHUs, or chillers can ripple through the entire schedule, monitoring serves as the early warning system, while control provides the steering action to keep the project on course.

Monitoring is only effective when the baseline plan is strong. Together, planning and monitoring form the backbone of project controlling. Too many projects drift because they rely solely on the baseline schedule prepared at the start. The reality is that the **planning–monitoring–controlling cycle must remain continuous until handover**. Strong monitoring systems don't just track dates; **they predict risks, highlight variances, resolve issues early, control costs, and ensure quality delivery.**

Golden Rule: *"Monitoring is not policing — it is enabling success."*

Projects that embed disciplined monitoring and proactive control consistently finish on time, within budget, and with fewer disputes, while those that don't are left firefighting and explaining overruns.

8.2.10 Monitoring & Controlling– At a Glance (Summary Note)

Key Takeaways:

- Monitoring = measuring performance; Controlling = taking corrective action.
- Variances must be tracked across **Schedule, Cost, Scope, and Quality**.
- EVM provides objective measures: PV, EV, AC → SV, CV, SPI, CPI.

- Monitoring must extend to **Procurement, Communication, Resources, Integration, Commercials, Compliance, and HSE.**
- **Tools:** Issue log, risk register, trackers, dashboards (offline/online).
- Dashboards act as the "car dashboard" for projects.
- Success lies in **updating data regularly, focusing on the critical path, and escalating early.**
- A project that is monitored well is controlled well.

Annexure 1: Earned Value Management

In HVAC projects, simply tracking "% complete" is not enough. A duct may be 80% installed, but if the cost spent is 120% of the budget, the project is already in trouble. To bring objectivity, we use **Earned Value Management (EVM)** — a technique that integrates **scope, schedule, and cost** into a single performance measurement framework.

EVM allows the project manager to answer three critical questions at any point in time:

1. **Are we on schedule?**
2. **Are we on budget?**
3. **What will be the likely outcome if current trends continue?**

Key Terms in EVM

- **Planned Value (PV):**

  The budgeted cost of work scheduled up to a point in time.
  Example: By week 4, 500 m^2 of ducting was planned at a cost of ₹5 lakh.

- **Earned Value (EV):**

  The budgeted cost of work actually completed by that time.
  Example: By week 4, 400 m^2 of ducting was installed; budgeted cost for this work = ₹4 lakh.

- **Actual Cost (AC):**

  The actual money spent on the work completed.
  Example: ₹4.5 lakh was actually spent to install 400 m^2 ducting.

Variance & Performance Indicators

From these three values, we derive:

- **Schedule Variance (SV) = EV – PV**
  Positive = ahead of schedule; Negative = behind.
  Example: EV (₹4 lakh) – PV (₹5 lakh) = –₹1 lakh → Behind schedule.

- **Cost Variance (CV) = EV – AC**
  Positive = under budget; Negative = over budget.
  Example: EV (₹4 lakh) – AC (₹4.5 lakh) = –₹0.5 lakh → Over budget.

- **Schedule Performance Index (SPI) = EV / PV**
  Ratio of progress achieved vs planned.
  SPI = 0.8 → only 80% of planned progress achieved.

- **Cost Performance Index (CPI) = EV / AC**
  Ratio of value earned vs cost incurred.
  CPI = 0.89 → spending more than budgeted for the work done.

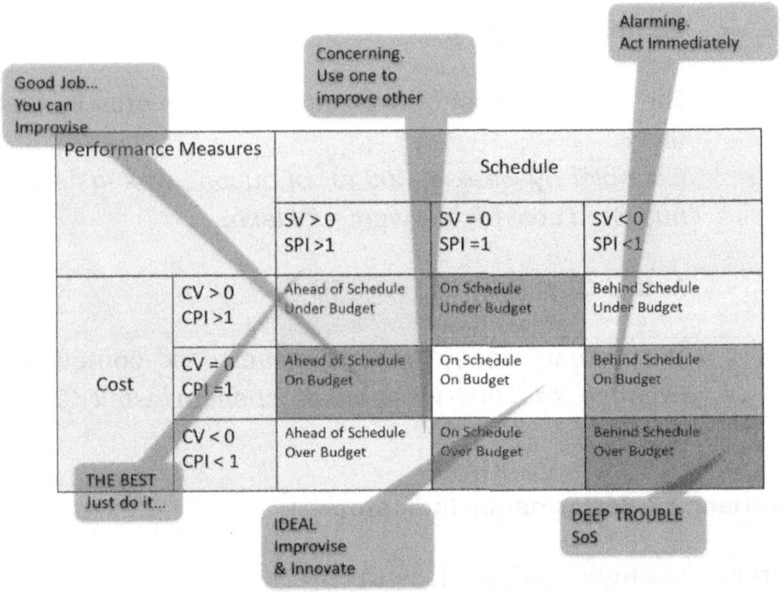

Figure 8.1 Earned Value Management Analysis

Mini EVM Example — Ducting Installation (step-by-step)

Scenario (by the end of Week 4)

- Planned work by Week 4: **600 m** of ducting at **₹1,000/ m**
- Actual installed by Week 4: **480 m**

- Actual cost spent to date: **₹520,000**
- Total budget for the ducting scope (BAC): **₹1,500,000**

Calculations (digit-by-digit)

1. **Planned Value (PV)** = Planned quantity × unit rate
   PV = 600 × 1,000 = **₹600,000**

2. **Earned Value (EV)** = Actual quantity installed × unit rate
   EV = 480 × 1,000 = **₹480,000**

3. **Actual Cost (AC)** = Actual money spent to achieve the EV
   AC = **₹520,000**

4. **Schedule Variance (SV)** = EV − PV
   SV = 480,000 − 600,000 = **−₹120,000** (negative → behind schedule)

5. **Cost Variance (CV)** = EV − AC
   CV = 480,000 − 520,000 = **−₹40,000** (negative → over budget)

6. **Schedule Performance Index (SPI)** = EV / PV
   SPI = 480,000 / 600,000 = **0.80**

7. **Cost Performance Index (CPI)** = EV / AC
   CPI = 480,000 / 520,000 = **0.9231** (≈ **0.92**)

8. **Estimate at Completion (EAC)** — simple method: EAC = BAC / CPI
   EAC = 1,500,000 / 0.9231 = **₹1,625,000**
   → Forecasted overrun ≈ **₹125,000** (EAC − BAC)

Quick table

Table 8.3 Variance & Performance Indicators

Metric	Formula	Value
PV	600 × 1,000	₹600,000
EV	480 × 1,000	₹480,000
AC	—	₹520,000
SV	EV − PV	−₹120,000
CV	EV − AC	−₹40,000
SPI	EV / PV	0.80
CPI	EV / AC	0.9231 (~0.92)
BAC	—	₹1,500,000
EAC (BAC/CPI)	BAC / CPI	₹1,625,000
Forecast Overrun	EAC − BAC	₹125,000

Interpretation (what this tells you)

- **SPI = 0.80** → The project is **behind schedule** (only 80% of planned work done).
- **CPI = 0.92** → project is **over budget** (you're getting ~₹0.92 of earned value for every ₹1 spent).
- **SV & CV negative** → immediate attention needed; otherwise delays & cost will compound.
- **EAC > BAC** → if current performance continues, expect roughly **₹125k** overrun on the ducting scope.

Recommended immediate actions (practical)

1. **Verify data accuracy**

- Confirm installed quantities (joint measurement), validate invoices, and confirm AC entries. EVM is only as good as your data.

2. **Diagnose root cause**

- Why behind schedule? (missing approvals, material shortage, manpower shortfall, rework)

- Why over budget? (overtime, rework, wastage, incorrect estimating)

3. **Short-term recovery (schedule)**

- Re-sequence non-critical activities.

- Add crews or shifts (only if cost-effective).

- Mobilise prefabrication or outsource parts of the work.

- Use CCPM feeding buffers if resource constrained.

4. **Cost containment (near-term)**

- Stop non-value activities; control overtime; reduce waste; renegotiate minor supplier rates.

- Record and classify rework costs separately for claims if scope/time changes caused it.

5. **Update forecast & escalate**

- Update EAC and submit impact note to CCB / client if additional budget/time is required.
- Propose a corrective recovery plan with a timeline and cost impact.

6. **Prevent recurrence**

- Improve QA at source (e.g., enforce fabrication checklists), train crews to reduce rework, tighten procurement follow-ups.

Practical notes for HVAC readers

- **Measure EV physically** (installed length, area, or completed deliverable) rather than using subjective "% complete". For ducting, physical measurement (m) is ideal.
- **Unit rates** must be consistent (use the BOQ/unit rate or agreed contract rates).
- **EVM is a management tool, not accounting** — use averaged resource rates for monitoring; accounting will reconcile actuals later.
- Use EVM **for decision-making**: SPI/<1 or CPI/<1 are triggers for investigation and corrective action thresholds (e.g., SPI < 0.9 or CPI < 0.95 → immediate review).

Table 8.4 Variances in a Nutshell

Variance Type	Cost Variance (CV)	Schedule Variance (SV)
Definition	The difference between budgeted and actual costs for work completed.	The difference between planned progress and actual progress at a given time.
Formula (where applicable)	CV = EV – AC CPI = EV / AC	SV = EV – PV SPI = EV / PV
HVAC Example	Overordering insulation → excess inventory → wastage and higher costs.	Duct fabrication drawings approved late → duct installation delayed → false ceiling works postponed.
Control Measures	- Strict procurement monitoring - Compare actual vs. BOQ regularly - Record/approve rework costs separately.	- Re-sequence non-dependent tasks - Increase manpower/shifts - Escalate bottlenecks (site not ready, other trades).

Quality Variance	Scope Variance
Non-conformance of work with specifications/standards.	Deviations between agreed scope and what is delivered.
No formula; tracked through NCRs and inspections.	No direct formula; measured against BOQ or baseline scope.
Welding by untrained worker → joint failure → rework and delays.	Client requests additional air outlet after drawings frozen → duct extension + diffuser + rebalancing.
- Stage-wise checklists - Supervisor/technician training - Empower QA/QC to reject work.	- Formal CCB process - Record impact on time/cost before approval - Communicate changes to stakeholders.

Table 8.5 Monitoring other key areas

Area	What to Monitor	How to Monitor	HVAC Example
Procurement	Status of long-lead items (POs, FAT, shipment, customs, delivery)	Procurement tracker, vendor progress reports, joint review meetings	Delay in cooling tower shipment identified early, re-sequenced ducting activities
Communication	Flow of information, meeting cadence, and version control of drawings	RACI matrix, minutes of meeting, drawing trackers	Avoid the wrong drawing version being used at the site
Resources	Availability and Utilisation of manpower, tools, and equipment	Daily manpower logs, equipment usage logs, resource histograms	Shortage of welders flagged → additional crew mobilised
Integration	Dependencies across trades (civil, electrical, plumbing, interiors)	Integrated schedule, BIM coordination, weekly coordination meetings	HVAC duct clashing with the electrical cable tray was detected in BIM, and was resolved before site execution

Commercials	Accounts Receivable (AR): client payments, Accounts Payable (AP): vendor payments, Cash flow curves (S-curve inflow vs outflow)	Invoice tracker, payment status reports, and financial dashboards	Client delays ₹1.2 Cr invoice → flagged early to avoid vendor payment default
Regulatory Compliance	Adherence to codes & approvals (NBC, ISHRAE, fire safety, statutory permits)	Compliance checklist, approvals register, consultant audits	Fire NOC pending → escalated to avoid occupancy certificate delay
HSE (Health, Safety & Environment	Safety adherence (PPE usage, permits-to-work, incident logs)	Daily toolbox talks, HSE checklists, audit reports	Unsafe scaffold detected → work stopped until corrected

Chapter 9:
Contracting Essentials for Project Managers

"The contract is not just a piece of paper — it is the roadmap that defines trust, clarity, and fairness in a project."

Learning Objectives

By the end of this chapter, the reader will be able to:

1. Understand the basics of contracting in HVAC projects.
2. Differentiate between contract types used in the Indian HVAC industry.
3. Recognise the importance of clear contract clauses (scope, BOQ, timelines, quality standards, payments).
4. Learn best practices in subcontracting and avoid common pitfalls.
5. Appreciate the role of technical, commercial, and legal clarity in project success.
6. Apply practical templates (Work Orders, BOQ reconciliation, closure checklist).

9.1 Introduction

Why Contracts Matter in HVAC

In India, the HVAC industry is mostly run by technocrats. While they excel at technical solutions, only a few have strong expertise in **contractual conditions, clauses, and legal implications.** As a result, contracts are often misunderstood as mere "legal paperwork."

In reality, a contract is the **operating manual of the project** — it defines scope, cost, timelines, responsibilities, risks, and quality expectations.

Unfortunately, in India, contracting practices are often weak:

- BOQ ambiguities cause endless disputes.
- Contractors and subcontractors rely on oral agreements.
- Variations go undocumented, leading to withheld payments.
- Contract closure is treated as an afterthought.

A poorly written or poorly executed contract can **erode trust, delay closing, and damage relationships.** On the other hand, a clear, fair, and well-managed contract protects both owners and contractors — ensuring smooth delivery and client satisfaction.

Golden Rule: *"If it is not written in the contract, it does not exist."*

9.2 Types of Contracts in HVAC Projects

Different types of contracts are used depending on the project's size, complexity, and approach to risk sharing.

a. Lump Sum / Turnkey Contracts

- **Fixed price** for completing the entire project.
- Suitable when the scope and specifications are very clear.
- Common in VRF, chilled water, or packaged AC jobs.
- **Risks:** If scope changes, disputes are inevitable. The owner benefits, but contractors carry high risk.
- **Example:** Military engineering services and certain government projects where the scope is well-documented.

b. Item Rate Contracts

Most common in HVAC projects involving supply, installation, testing, and commissioning.

- Contract includes technical specifications and commercial requirements.
- BOQ and approved makes are provided by the consultant.
- **Payments:** Based on BOQ and actual joint measurements.
- Widely used for SITC of chiller plants, VRF systems, or packaged units with ducting, piping, and insulation.

- **Risks:**
    - Disputes if the measurement methodology isn't clearly defined.
    - Warranty responsibilities and system performance validation are often contested.

c. EPC (Engineering, Procurement, Construction) Contracts

Common in large construction projects where HVAC is part of a bigger package.

- Contractor responsible for **design, procurement, and execution.**
- Effectively a lump sum contract covering SITC.
- **Risk:** Contractor bears most risks (design, cost, and schedule).
- **Example:** Airports, metros, large infrastructure projects.

d. Design & Build Contracts

A hybrid where the contractor is responsible for **design + SITC + handover.**

- Contractor's design team prepares Basis of Design, drawings, BOQ, and commercial terms.
- Single party accountable for both design and execution.
- Gaining popularity in India for complex commercial projects.
- **Risk:** Contractor owns both design and delivery risk.
- **Advantage:** Faster decision-making and single-point accountability.

e. Rate Contracts / Annual Maintenance Contracts (AMC)

- Common for off-the-shelf products like split units or VRF systems.
- Long-term agreements for **operation and maintenance** after handover.
- Widely used by government and large organisations, often directly with OEMs or their dealers.
- **Risk:** Scope must be clearly defined (parts, labour, consumables, response time).
- **Example:** Corporate offices signing 3–5-year AMC for VRF systems with OEM service providers.

Tip: *Select the contract type based on clarity of scope, project complexity, and risk appetite. Poor contract selection at the start often leads to disputes later.*

9.3 Key Clauses in HVAC Contracts

Key Clauses in HVAC Contracts — structure + practical examples

Before you start:

Every member of the project team must read the contract in full *before* mobilization. Annotate the clauses, write down the operational impact, and clarify ambiguous terminology with the project's senior contracts/legal owner. Wrong assumptions about clauses lead to penalties, withheld payments, and bitter disputes.

1. Scope of Work

What it covers: detailed scope, inclusions, exclusions, drawings, measurement methods, and test procedures. **Practical examples/pitfalls:**

- If **"ducting"** in scope does not specify whether grilles/diffusers are included, disputes arise at measurement. Bends are measured differently from straight pieces.
- If **upstream civils** (wall openings, chases) is not listed as client scope, site delays become the contractor's responsibility.
  Action: Create a Scope Clarification Log, mapping each major BOQ line to: drawing ref / in-scope / out-of-scope / owner responsibility.

Golden Rule #1: *If it is not written in the contract, it does not exist.*

2. Technical Specifications (Standards)

What it covers: standards the work must meet (ISHRAE/ASHRAE, SMACNA, IS codes, ECBC, UL, etc.). **Example:** Duct fabrication quality must meet SMACNA; leakage class must be specified. If the contract just says "SMACNA" without a leakage class, you'll get differing interpretations.
Action: Prepare a Spec Compliance Matrix (clause → expected value → evidence required, e.g., FAT, test certificates).

3. BOQ & Measurement Methodology

What it covers: unit descriptions, units of measurement, measurement method (centreline, surface area, run length), frequency of joint measurements.

Practical examples:

- Pipe: is measurement by centreline length or external perimeter? That changes quantities significantly.
- Insulation wastage % not specified → disputes on acceptance of measured quantities.
- **Action:** Include a Measurement Protocol annexure: show sample joint measurement procedure, required signatories, and timing (e.g., measure before false ceiling closure).

Golden Rule #2: *Never execute a variation without written approval.*

4. Payment Terms

What it covers: mobilisation advance, interim bills, milestone payments, retention, BGs, and release triggers.

Practical examples:

- Mobilisation 10% on PO, 50% of material supplied on site as advance.

- Retention 5–10% held till practical completion; retention release after DLP only on submission of all warranty docs.
- **Risk:** Stringent client-favoured terms (e.g., payment only after consultant certification) can break contractor cash flow.
- **Action:** Build a Cashflow Model from the contract payment schedule and compare it to the project burn rate. Escalate the funding gap to management early.

> **Golden Rule #3:** *Cashflow kills projects faster than delays — study the payment terms like you study the drawings.*

5. Timelines & Liquidated Damages (LD)

What it covers: completion dates, milestone dates, LD rates and cap, definition of "delay", and force majeure.

Practical examples:

- LD phrase A: "0.5% per week on the **delayed portion**" vs phrase B: "0.5% per week on the **contract value**". These are drastically different in impact.
- Cap example: LD capped at 10% of contract value. **Action:** Simulate LD exposure for plausible delay scenarios. Negotiate caps/relief for client delays (late drawings, site readiness).

> **Golden Rule #4:** *LD clauses may look harmless — simulate the impact.*

6. Warranty & Defects Liability Period (DLP)

What it covers: warranty start date, what's covered, response SLAs, spare parts, and test certificates.

Practical examples:

- Some vendors count warranty from the date of supply; you need a warranty from the commission date or handover.
- Commissioning-related performance guarantees (e.g., chiller COP) must have acceptance criteria and a measurement method.

Action: Prepare a Warranty File during execution — test certs, FAT reports, serial numbers — so handover is clean, and retention is released.

Golden Rule #5: *Warranties must start from commissioning, not supply.*

7. HSE Requirements

What it covers: PPE, permits to work, HSE reporting, zero tolerance clauses, and penalties for noncompliance. **Practical examples:** A zero-tolerance clause may require immediate site suspension for an HSE breach. That can cost more than LD.

Action: Ensure the HSE plan and toolbox talk schedule are part of the contract deliverables and assign HSE responsibility in the staffing plan.

8. Price Variation / Escalation Clauses

What it covers: how material price changes (copper, steel) and statutory changes (GST / import duties) are handled. **Practical examples:** Copper price surge formula tied to LME index or a fixed %; if missing, contractor bears the risk. **Action:** If project > 6–9 months, negotiate an escalation formula or a periodic re-pricing trigger. Capture index, base month, and calculation method in writing.

> **Golden Rule #6:** *Escalation clauses are insurance against copper/steel price shocks — negotiate them early.*

9. Dispute Resolution

What it covers: escalation matrix, consultant role, adjudication, arbitration, jurisdiction, and timelines.

Practical examples: Without an escalation ladder, small issues go straight to formal dispute — costly and slow.

Arbitration location and law (e.g., arbitration in India under the Indian Arbitration Act) matter.

Action: Ensure a staged dispute process: Site/PM → Contracts Manager → Project Director → Conciliation/Arbitration (with timelines for each step).

10. Termination Clause

What it covers: termination for convenience vs for cause, notice, settlement, demobilisation costs, payment for work done.
Practical examples: "Termination for convenience" may only

pay for completed work minus overheads unless an exit formula is defined.

Action: Negotiate fair termination compensation and include demobilisation cost recovery if termination is not the contractor's fault.

9.4 Contract Highlights

The following activities are to be highlighted as soon as one gets the contract.

1. **Clause Impact Log** (Clause → Operational Impact → Owner / Action / Risk rating).
    - Example entry: Payment Terms → 30-day payment cycle → Cashflow shortfall → Finance & PM to escalate.
2. **Joint Measurement Procedure** uploaded to site and enforced (who signs, when).
3. **Variation Order SOP** — do not execute variations without written approval and cost/time impact.
4. **Contract Review Session** — PM + Contracts Manager + Finance + QA + Procurement must meet and confirm interpretations. Document minutes.
5. **Legal Review** for high-risk clauses (LD, termination, indemnities, arbitration). Don't let legal be an afterthought.
6. **Document Control & Single Source of Truth** — link contract clauses to drawings, submittals, and the BOQ in a Contract Traceability Matrix.

9.4.1 Practical examples — quick scenarios

- **Example A — Duct measurement dispute:** Contract is silent on measurement norm. Contractor used centreline length; client used external length → 8% variation and withheld payment. *Prevention:* Define the measurement method up front and insert example calculations in the contract annexure.

- **Example B — Warranty start date argument:** Supplier claims warranty from supply date; contractor claims from commissioning. The client withholds retention till the warranty certificate for the commissioned date is provided. *Prevention:* Define warranty start event: "date of successful performance test & acceptance."

- **Example C — Customs delay & price escalation:** Imported AHU held in customs for 6 weeks; prices rise and delivery is delayed. If the contract has no escalation clause or delivery relief, the contractor bears the cost. *Prevention:* Include customs/delivery force majeure and escalation formula.

9.4.2 PM checklist on receiving contract

1. Read the contract fully — once is not enough. Annotate and flag unclear words.

2. Create a Key Clause Summary (1-page): Scope, Payment, LD, Warranty, DLP, HSE, Variations, Dispute route. Share with the team.

3. Run a contract impact workshop with Finance, Procurement, QA and Site.

4. Build short simulations: cash flow, LD exposure, variation cost sensitivity.

5. Update project risk register with contract-derived risks.

6. Keep the legal in the loop for any unusual indemnities or onerous clauses.

NOTE:

This chapter focused on minimal practical contract management steps and examples only. For complex or high-value, detailed contractual language — especially indemnities, arbitration clauses, or international law — get qualified legal advice. Contracts are enforceable legal documents and small wording changes can have big financial impacts.

9.5 Subcontracting in HVAC Projects

Most HVAC contractors do not execute all works directly. Specialised subcontractors handle critical parts of the project, such as:

- **Duct fabrication and erection**
- **Piping and welding**
- **Insulation works**
- **Minor civil works**
- **Electrical cabling and panels**
- **BMS integration and controls**

Risks of Informal Subcontracting)

Unfortunately, many subcontracting arrangements in India are **informal**, relying only on oral instructions and trust. This often leads to:

- No **written work order**, only verbal agreements.
- No clarity on **quality standards** or scope.
- No **formal agreement** on quantities, rates, or timelines.
- Poor adherence to **safety protocols**.
- Payment disputes due to a **lack of measurement methodology**.

Best Practice: Formal Work Orders

To avoid disputes, issue a **formal Work Order** to every subcontractor **before work begins**. This must clearly state:

- **Scope of work** and deliverables.
- **Agreed rates and quantities**.
- **Measurement methods** (duct area, pipe length, insulation thickness).
- **Applicable technical specifications and standards**.
- **Responsibility for tools, tackles, scaffolding, consumables, and transport**.
- **Safety compliance requirements** (PPE, permits, HSE standards).

- **Payment terms**, retention, and penalties for delay or poor quality.

> **Golden Rule:** *"Subcontractors are your partners, not just labour suppliers. Clear agreements protect both sides."*

Practical Tip

Respect + Clarity = Performance

A subcontractor who feels respected and has clarity on scope will often deliver better quality and be more loyal to your projects than one managed by fear or ambiguity.

9.6. Case Study

Background:

A hospital HVAC project in India involved multiple subcontractors. Ducting was subcontracted without formal work orders, based only on the site engineer's oral instructions.

What Went Wrong:

- The client disputed BOQ measurements.
- Excess materials and wastage due to rerouting were never documented.
- Subcontractor demanded higher payments, but no work order existed to support his claim.
- Client withheld 15% of final payments citing "incomplete documentation."

Outcome:

The project closing dragged on for 8 months. The subcontractor suffered losses. The contractor's relationship with the client soured.

Lesson:

Formal agreements and documentation are not bureaucracy — they are insurance against disputes.

9.7 Conclusion: Why Contracting Basics Matter

Contracts in HVAC are not legal formalities — they are the **operating manual of the project**. A well-structured contract, supported by transparent subcontracting practices, minimises disputes, ensures fair risk sharing, and keeps the project aligned to scope, cost, and time.

In India, where informal agreements are common, project managers who insist on **written clarity, structured clauses, and professional subcontracting** stand apart. They avoid last-minute disputes, protect margins, and most importantly, hand over projects with confidence and credibility.

Contract management in HVAC projects is not about reading fine print in isolation; it is about **connecting the contract to daily project decisions.**

- Every delay, dispute, or withheld payment can usually be traced back to an overlooked clause or assumption.
- A well-understood contract protects both owner and contractor, reduces surprises, and ensures smoother project delivery.
- Project managers must treat contracts as their **operating manual** — knowing which clauses affect

scope, cashflow, timelines, risk allocation, and warranties.

- In the Indian HVAC context, where oral agreements and undocumented variations are still common, written clarity is the strongest shield.

"A strong contract doesn't just end disputes — it prevents them."

Chapter 10: Critical Chain Project Management (CCPM)

"Buffers don't add time; they protect the promise." — Dr. Eliyahu M. Goldratt

Key Takeaways

Why Critical Chain Project Management is Necessary for HVAC Projects

By the end of this chapter, the reader will be able to:

Recognise why HVAC projects require critical chain project management.

- How to identify bottlenecks in HVAC Project Management Lifecycle.
- Understand the scheduling user constraints-based planning
- Monitor projects using buffer management techniques
- Use the lifecycle as a roadmap to deliver projects on time, within budget, and to quality standards consistently

10.1 Introduction:

We have discussed in previous chapters various challenges faced by HVA project managers. These include resource non-availability, demanding, at times, unreasonable timelines. Over promise. Under delivery situations and even padded durations. Even though the project manager plans the project very judiciously, incorporating all the risks, schedule overruns, and cost overruns derail the project. This is a serious situation that occurs in over 30% of the projects.

Chain Project Management (CCPM), derived from the Theory of Constraints (TOC), addresses these challenges by focusing on resource constraints, eliminating harmful behaviours like bad multitasking and student syndrome, and introducing explicit schedule protection via buffers. This chapter explains CCPM in the context of HVAC delivery, complete with practical examples, buffer concepts, monitoring, and a realistic case study.

10.2 Basic concepts of Critical Chain project management.

What is "Critical Chain"?

The first question is what Critical Chain is and how different it is from critical path management (CPM), which most managers are familiar with. As we all know, **the Critical path focuses on task precedence and durations and alerts the users, any deviation in these tasks impacts the delivery of the project**. Noncritical path tasks go unnoticed. Eventually, entire project tasks become critical and managing all "critical" tasks leads to firefighting mode.

However, **Critical Chain adds resource constraints to precedence logic, identifying the sequence of tasks constrained by both dependencies and shared resources** (e.g., a single insulation team shared across multiple zones, one TAB team servicing several floors, one supervisor managing different zones of a multi-storied commercial HVAC project).

The plan is then protected with time buffers (project and feeding buffers) and managed using buffer consumption rather than task-level due dates.

10.3 CCPM Vs Improve HVAC project execution efficiency.

HVAC projects have constraints that potentially impact the project schedules and costs. HVAC completion is often on the project's critical path to occupancy—slippage is costly, making robust schedule protection and responsive monitoring essential. There are resource constraints: Typical constraints include crane time, Site availability, Adequate power availability at the site, ducting erection team, certified welders, BIM coordination hours, and commissioning team availability.

In addition, variability is also significant in areas like site access limits, design clarifications, late submittal approvals, long-lead time equipment, actual installed conditions and payment-related issues.

10.3.1 Harmful behaviours of the system

CCPM achieves consistent project delivery by eliminating certain inherent harmful behaviours of the system. A brief description of those behaviours is provided below:

a. **Multi-Tasking:** It is a very common practice in HVAC to assign a set of resources to multiple tasks of the same project or to different projects in each period. This is due to a shortage of resources or expensive resources. Frequent task switching reduces throughput, increases errors, and elongates cycle time. Work-in-progress (WIP) balloons and downstream teams are idle or resequencing, compounding delays.
b. **Student Syndrome (Starting Late):** Why do projects fail even though the timelines are padded with safety times? As the assigned resources are aware that there is an element of safety available, the task is started late. This results in a non-critical task turning red into a critical task. In other words, the tendency to defer effort until late in the allotted time eliminates natural slack, allowing minor issues to compromise delivery.

c. **Parkinson's Law (Work expands to fill the time available):** When a task is completed much before the scheduled time, the task is not reported as completed. It is stretched till the due date. For example, the duct fabrication shop uses the full scheduled time even when actual content could be turned around faster, consuming scarce shop capacity and constraining downstream installation.

10.4 CCPM Philosophy:

Critical chain project management is based on the philosophy of "Theory of Constraints" proposed and popularised by Dr Eliyahu Goldratt.

CCPM focuses on shorter, aggressive-but-realistic task durations; forbidding individual task safety; centralised buffers; explicit WIP control; resource-levelling that respects the constraint; strict focus to stop multitasking; and buffer-based execution management. It is a framework, a philosophy that identifies the project constraints, questions the assumptions and eliminates the project dilemmas by injecting an alternate solution.

10.4.1 Buffers in CCPM:

CCPM Philosophy is to eliminate the "inbuilt safety lead times" and replace it with more efficient and visible Project and Feeding Buffers.

1. What is a Project Buffer?

- **Definition:** A single schedule buffer placed at the end of the critical chain, just before the project's completion or handover milestone.

- **Purpose:** Absorb variation in the critical chain so the overall delivery date is protected.

- **Sizing:** The Sizing method used in industry is by totalling the removed task-level "safeties" on the critical chain and converting a portion (e.g., 50%) into a project buffer or using statistical/quantile approaches where data is available.

- **Application:** In HVAC projects, this protects against late equipment delivery, extended installation time, surprise rework in ductwork clearances, or delay in commissioning and TAB activities completion.

2. What is a Feeding Buffer?

- **Definition**: Feeding Buffers are placed where non-critical (feeding) chains join the critical chain.

- **Purpose:** Prevent delays from non-critical work

- **Sizing:** The Size of the feeding buffer is shorter than the project buffer, based on variability and the length/uncertainty of the feeding chain segment.

- **Application:** For example, wall openings for duct connections, piping installation clearance, Scaffolding /access platform fabrication) from propagating into the critical chain.

3. What is a Resource Buffer?

- **Definition**: Critical Resource Buffers are kept in standby mode.
- **Purpose:** Prevent delays from critical activities in case of an emergency.
- **Sizing:** The Size of the Resource buffer is decided based on the application and its impact.
- **Application:** For example, staging a backup crane day, pre-booking TAB personnel, reserving overtime approval, or pre-qualifying alternate duct shop capacity. Note, these are not time buffers, but alerts/preparations to ensure a critical resource is ready when needed.

10.5 CCPM Process examples for HVAC Applications.

Commissioning/TAB:

- **Bad Multitasking**: TAB team is split thin across many floors to "show progress," leading to retesting and repeated partial visits. **Resolution:** One-flowline approach—finish Level 3 completely before Level 4; protect with a feeding buffer ahead of final integrated testing.
- **Student Syndrome**: Starting point-to-point commissioning checks late because the prior installation tasks had nominal slack; then discovering mislabelled terminations triggers rework with no time left. **Resolution:** earlier pre-commissioning and readiness checks, add resource buffer.

10.6 Project Monitoring for CCPM application

CCPM reframes control from task deadlines to buffer consumption. Rather than pressuring each activity to hit padded dates, the team monitors how much of each buffer has been used relative to work completed and reacts early. Figure 10.1 Buffer Monitoring in CCPM given below shows the Time vs Buffer consumption using Fever chart or heat map.

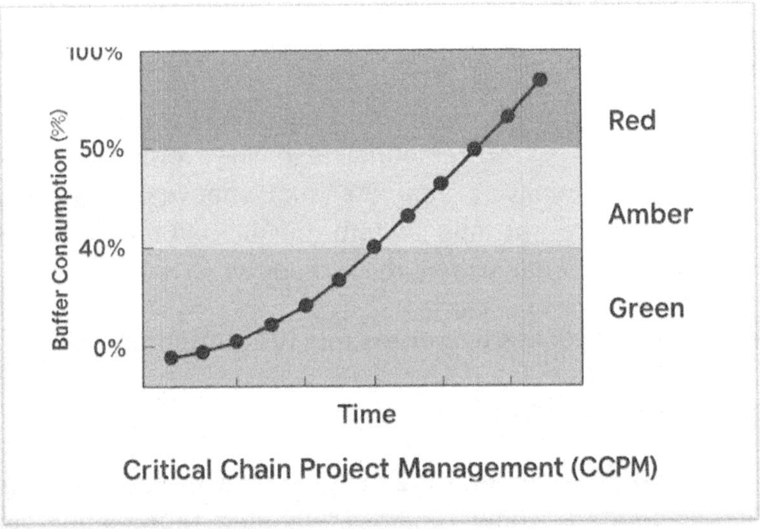

Figure 10.1 Buffer Monitoring in CCPM: Fever Chart

Key elements:

10.6.1 Fever Chart (Project Buffer):

- X-axis: % of time critical chain completed.
- Y-axis: % of project buffer consumed.

Zones:

- **Green:** Buffer consumption ≤ progress; continue focused execution.

- **Amber:** Buffer consumption approaching progress; trigger corrective actions (expedite submittals, re-sequence, allocate additional crew).

- **Red:** Buffer consumption exceeds progress; escalate, invoke contingency (e.g., alternative vendor, overtime, additional crane day).

10.6.2 Feeding Buffer Tracking:

- Monitor each feeder's burn rate vs. its chain progress; if a feeder buffer is burning faster than the feeder is completing, advance mitigation before it jeopardises the critical chain.

10.6.3 Daily/Weekly Constraint Reviews:

- Confirm the constraint (e.g., TAB team/ Piping team) is fully utilised on the highest-priority task, not starved or blocked by missing pre-requisites.

- Protect the constraint: pre-assemble materials, clear access, and complete QA/QC checklists ahead.

- Prepare the pre-Commissioning checklist right at the planning stage and identify the resources required.

10.6.4 WIP Limits and Task Focus

- Enforce strict policies against parallel starts beyond capacity.

- Maintain a visible queue for the constraint; do not overload upstream if it creates unfinishable inventory.

10.6.5 Short Interval Control with CCPM lens:

- 24–72-hour look-ahead tied to buffer status.
- Daily standups: What tasks are "ready-to-go"?
- What is blocking the constraint?
- Which buffer is at risk?
- What action today reduces buffer burn?

10.6.6 Rapid Escalation:

- Predetermine response ladders for amber/red signals: vendor escalation, overtime, alternate supplier, engineering clarifications, or temporary redundancy (e.g., a second rigging crew for a short burst).

By applying these CCPM practices deliberately, HVAC teams move from reactive firefighting to proactive control, delivering more reliable outcomes with less hidden stress on crews and budgets.

10.7 Summary:

- ➢ CCPM reframes the schedule around resource constraints and replaces scattered task safety with centralised buffers that protect delivery.
- ➢ Project buffers shield the contractual date; feeding buffers prevent non-critical delays from infecting the critical chain.
- ➢ Eliminating bad multitasking and student syndrome is not optional—WIP limits, readiness checks, and focused work are central to CCPM.

- Monitoring is buffer-centric: act when buffer burnout paces progress; protect the constraint with preparedness and rapid escalation.

- In HVAC projects—where variability and resource contention are the norm—CCPM consistently improves reliability, reduces rework, and aligns teams around real priorities.

A Simple CCPM Checklist for HVAC Project Managers

- ✓ Network complete with resource calendars aligned.
- ✓ Aggressive durations agreed with foremen and vendors.
- ✓ Project buffer set at the end of the critical chain.
- ✓ Feeding buffers set at key merges (electrical → controls, fab → install, flushing → startup).
- ✓ Resource buffers/alerts placed for cranes, rigging, factory reps, TAB, BAS programming.
- ✓ WIP limits established; explicit "do not start" rules.
- ✓ Fever chart and buffer dashboards installed; weekly reviews booked.
- ✓ Readiness checklists for constraints; gatekeeping enforced.
- ✓ Escalation triggers are defined as amber/red buffer status.
- ✓ Post-mortem to adjust buffer sizing and readiness standards for next project.

Chapter 11: Project Closing

"Closing is not just about finishing — it's about handing over a system that performs as promised."

Learning Objectives

After reading this chapter, the reader will be able to:

1. Understand the importance of structured project closure in HVAC projects.
2. Identify key activities in project handover, including documentation, training, and commissioning validation.
3. Manage snag lists, warranty obligations, and closure approvals.
4. Recognise the role of client satisfaction and feedback in successful closure.
5. Apply practical tools like snag lists, handover checklists, and lessons learned templates.

11.1 Introduction

In HVAC projects, the final impression is often the lasting one. As PMBOK rightly explains, project closing is as important as project initiation. This phase provides a unique opportunity to step back and take a holistic view of both planning and execution, ensuring that the intended outcomes have been achieved.

A beautifully designed and executed system can still leave the client dissatisfied if the handover is rushed, incomplete, or poorly documented. Unfortunately, many HVAC projects stumble at this stage, with unresolved snags, missing documents, or O&M teams left without adequate training. Too often, project managers in the HVAC industry perceive closure only as a commercial activity. While contract and financial closure are indeed essential, without a proper technical closure that satisfies the client, commercial closure quickly becomes a nightmare for all stakeholders.

Proper project closure is therefore not just a contractual formality — it is an opportunity to:

- Secure client satisfaction.
- Protect against disputes.
- Ensure long-term system performance.
- Wrap up commercial and contractual obligations cleanly.
- Capture valuable lessons for future projects.

In simple terms:

> **"Projects don't end when the last screw is tightened — they end when the owner is confident to operate, maintain, and benefit from the system."**

11.1.1 When Does Project Closing Begin?

Before proceeding to the closure process, it is important to understand that closing is **triggered by specific events**:

✔ **Formal acceptance of project deliverables** by the sponsor or customer.

or

✔ **A decision to prematurely terminate** the project or a phase, either for cause or for convenience.

11.2 Project Closing Process Overview

The Reality of Closing in HVAC Projects

In the HVAC industry, ***most projects are completed but not truly closed***. This statement reflects the current state of understanding of the project closing process. Too often, closure is assumed to mean nothing more than submitting a few documents, running the plant, and showing that the conditioned areas are cooling.

Project closing is a systematic and professional process. It requires completing all pending project activities across technical, commercial, and administrative domains, and then comparing the planned objectives with the actual delivered outcomes. Only when this structured approach is followed can the project team — and the client — be confident that the project has truly met its purpose.

11.2.1. Technical Closing

The focus of **Technical Closing** is to ensure that the HVAC system performs as per the **Owner's Project Requirements (OPR)** and meets all contractual specifications. It is the foundation upon which contract and administrative closure can proceed. All technical-related activities are

Key Activities in Technical Closing

1. **Final Documentation**

    - All technical documents are updated and marked as *Final Versions*.
    - Submissions include:
        - Final product/service for which the project was undertaken.

    - Project acceptance form/certificate.
    - Final performance report as per contract documents, submitted by the HVAC contractor.
- Verified and approved by the consultant.
- Formally accepted by the client's team.

2. **Snag List Completion**
    - All snag items tracked to closure.
    - Priority: Complete before client occupancy — once the building is in use, arranging plant shutdowns to fix snags becomes difficult and costly.
3. **System Validation**
    - Testing, Adjusting, and Balancing (TAB) reports validated.
    - Functional Performance Testing of major equipment (e.g., chillers, AHUs, VRFs).
    - Seasonal testing conducted as per contract requirements.
    - **Formal Acceptance & Warranty**
        - Getting sign-off from consultant and client.
        - Defects liability period (DLP) obligations.
        - Warranty tracking and vendor obligations
4. **O&M Documentation**
    - Handover of O&M manuals, warranties, and as-built drawings.
    - O&M manual should include:

- Design intent and design brief.
- As-built drawings.
- Final calculations and equipment selection basis.
- Approved technical details and specifications.
- Photographs of plant rooms, equipment, pipes, ducts, and cables (with painting and labelling complete).
- Training records.
- Performance reports.
- Client handover checklist.

5. **Training & Knowledge Transfer**

- Training of the client's operations O&M team in system operation, safety protocols, and preventive maintenance.
- Demonstrating system operation, troubleshooting, and safety protocols.
- Importance of O&M manuals + system manuals.
- Training records documented as part of the handover package.

"Handover day should not be treated as a mere formality." It is the client's final impression of the project. A professional handover includes a structured demonstration of the system, a walkthrough of major plant areas, ceremonial signing of completion documents, and a clear explanation of O&M manuals. A well-managed handover builds confidence, leaves a positive memory, and often paves the way for repeat business.

Responsibilities

- Project Manager
- QA/QC Team
- Commissioning Lead

Table 11.1 Training & Knowledge Transfer

Activity	Responsible	Status (Open/In Progress/ Closed)	Remarks
All documents updated and marked as Final Version	PM / Documentation Lead	Open	
Final product/service submission	Contractor	In Progress	
Project Acceptance signed	Client / Consultant	Open	
Performance report submitted (as per contract)	Contractor	Closed	
Snag list prepared	QA/QC	Closed	
Snag list items resolved & verified	Subcontractors / Consultant	In Progress	
TAB (Testing, Adjusting, Balancing) completed	Commissioning Team	Closed	
Functional Performance Testing done (Chiller, AHU, VRF, etc.)	Commissioning Team	Closed	
Seasonal testing completed	QA/QC / Client	In Progress	
O&M manuals submitted	Contractor	Closed	
Warranties and guarantees submitted	Procurement/ PM	Open	
As-built drawings submitted	Design / Drafting Team	Closed	
Photographs of the plant room, ducts, pipes, labelling submitted	QA/QC	Closed	
Client O&M team training completed	Commissioning Lead	Closed	
Training records submitted	PM / Documentation Lead	Closed	

11.2.2. Contract Closure

The focus of Contract Closure is on settling all commercial and contractual obligations of the project. It ensures that financial matters, vendor obligations, and client commitments are completed cleanly, leaving no open disputes or liabilities.

Key Activities in Contract Closure

1. **Final Measurement and BOQ Reconciliation**

    - Many conflicts arise between quantities supplied, installed, measured, and invoiced — often leading to disputes between the contractor, subcontractors, and client representatives.

    - To avoid this:

        - Maintain all Goods Received Notes (GRNs) for supply quantities.

        - Take joint measurements immediately after completion of each work section

Example: If ducting and insulation are completed and the false ceiling is closed, it becomes impossible to measure. Joint measurements during erection avoid disputes later.

        - Record wastage due to design changes, rerouting, or workmanship errors at the time of occurrence.

2. **Deviation and Variation Management**

    - Prepare a deviation statement for all variations between planned vs. actual quantities.

    - Obtain amendments for all deviations and additional works carried out.

  - *Tip:* Without a formal amendment, most finance departments do not accept excess quantities, non-standard items, or additional works (e.g., extra isolator, revised ducting).

3. **Invoice Reconciliation and Final Billing**

- Prepare a reconciliation statement with joint measurements and deviation statements.

- Raise the final invoice strictly as per contract payment terms.

- Ensure consultant and client representative approval for all quantities.

- If performance bank guarantees are required, obtain the template well in advance and submit it along with the final bill.

4. **Retention Release**

- Apply for retention release after the Defects Liability Period (DLP) as per contract.

5. **Claims and Dispute Settlement**

- Resolve pending claims or disputes formally before closure.

6. **Subcontractor Contract Closure**

- In HVAC projects, many subcontractors unfortunately work without formal work orders or agreements, often based only on oral instructions from site engineers. This leads to disputes when engineers move on, and records are lost.

- To prevent this:
    - *Issue work orders or amendment orders for all subcontractor works before final billing.*
    - *This ensures clarity on scope, quantity, quality, and pricing, and helps finance teams* provision correctly for payments and warranty services.

7. Warranties and Guarantees

- Hand over all warranty/guarantee documents to the client.
- Challenge: Test certificates and warranties are often lost in the time gap between delivery, installation, and commissioning.

Tip: Maintain a dedicated file (physical + digital) for all warranty documents from the start. Missing warranties often lead to large amounts being withheld at final settlement.

Responsibility: Commercial Manager, Contracts Manager, Project Manager.

Golden Rule: *"Never delay joint measurements and documentation — today's small negligence becomes tomorrow's big dispute."*

Table 11.2 Contract closing checklist

Activity	Responsible	Status	Remarks
Final BOQ reconciliation completed	Contracts Manager		
Final measurements jointly verified & approved (before false ceiling or concealment)	Site Engineer / Consultant		
GRNs (Goods Received Notes) tallied with invoices and installation quantities	Storekeeper / Procurement		
Deviation statement prepared (planned vs. actual)	PM / Contracts Manager		
Contract amendments/variations approved (extra quantities, non-standard items, rework)	Contracts / Client Rep		
Subcontractor work orders/amendments issued (avoid oral-only instructions)	PM / Site Engineer		
Subcontractor invoices reconciled & cleared	Commercial / PM		
Final invoice prepared with reconciliation statement & deviation approvals	Contractor / PM		
Performance bank guarantee template obtained & submitted	Finance / PM		
Retention release applied (post DLP, as per contract)	PM / Finance		
Claims/disputes documented and resolved	Contracts Manager		
Warranties & guarantees compiled in a single file (test	QA/QC / Documentation		

certificates, warranty cards, guarantees)			
Warranty documents handed over to client (signed acknowledgement)	PM / Client Rep		
All debris has been cleared from the site	Site Supervisor / Contractor		
Temporary offices and facilities dismantled/removed	Site Admin / Contractor		

11.2.3. Administrative Closure

While contract closure ensures financial and legal obligations are settled, **administrative closure ensures that knowledge, records, and lessons are preserved for the organisation's future benefit**.

In many HVAC projects, teams discover highly effective solutions to recurring site challenges — innovative methods that save time and money. Unfortunately, without a structured process for **knowledge retention and sharing**, these solutions remain with senior engineers or get lost once the project closes. This leads to repeated mistakes, trial-and-error approaches, and wasted effort across projects.

Administrative closure is the step that prevents this cycle. It focuses on:

- **Knowledge management.**
- **Updating organisational assets.**
- **Formally documenting the project's outcomes.**

Organisational Process Asset Updates

1. **Project Documents**

    - Scope statement, schedule, technical submittals, cost reports, project calendars, risk registers, procurement logs, communication logs, and issue logs are archived.

2. **Project or Phase Closure Documents**

    - Formal documentation indicating project/phase completion.
    - Transfer of completed deliverables to the operations and maintenance (O&M) teams.

3. **Lessons Learned**

    - Documented and stored in the organisational knowledge base.
    - Shared across teams for use in future projects.
    - Preparing final project performance reports.
    - Updating organisational knowledge base.

11.2.4 Project Closing Meeting

The **Project Closing Meeting** is not a ceremonial step — it is the most valuable opportunity to consolidate learning, capture performance insights, and ensure organisational growth. By the end of the meeting, the project team and stakeholders should walk away with **clear, documented outcomes** that strengthen future projects.

- Design team
- Execution team
- Procurement team

- Commissioning team
- Logistics and supply chain
- Quality
- Maintenance team
- Commercial and contracting team
- Finance
- HR
- Senior/top management
- And other stakeholders as required

The **Project Closing Meeting** is not a formality — it is the single most important opportunity for the organisation *to capture lessons, celebrate achievements, and prepare for the future*. A well-run closing meeting ensures that the project's challenges and successes are not forgotten but are transformed into actionable improvements for the next project.

1. Objectives of the project Closing Meeting:

- Capturing project insights for future projects.
- Conducting a post-project review (successes + areas of improvement).
- Feeding data into the organisational knowledge base.

2. Key Outcomes of the Closing Meeting:

a. **Client & Customer Feedback**

- Present overall client satisfaction and feedback.
- Capture appreciation notes and unresolved concerns.
- Discuss system performance against Owner's Project Requirements (OPR).

b. **Team Recognition & Appreciation**

- Acknowledge contributions of design, execution, commissioning, procurement, and subcontracting teams.
- Share "above and beyond" efforts.
- Motivate teams by highlighting success stories.

c. **Lessons Learned – Structured Review**

- **Continue Doing**: Practices that worked well (e.g., early procurement of long-lead items).
- **Stop Doing**: Ineffective practices (e.g., relying on oral instructions for subcontractors).
- **Start Doing**: New practices to be adopted (e.g., digital snag trackers, cloud-based document sharing).

d. **Deviation Analysis**

- Review actual vs. planned outcomes in scope, cost, and schedule.
- Highlight major rework instances and causes.

- Discuss efficiency gaps (e.g., commissioning delays due to missing access panels).

e. **Risk & Procurement Review**

- Risks planned vs. risks that actually occurred.
- New risks that emerged during execution.
- Procurement audit: Vendor performance, delivery delays, import/customs issues.

f. **Quality & Safety Review**

- Quality successes (e.g., zero NCRs on duct insulation).
- Quality issues and preventive actions for future projects.
- Safety incidents, compliance review, and new safety tool requirements.

g. **Technology & Innovation Review**

- Tools and methods adopted (BIM, MS Project, mobile apps, testing tools).
- Evaluate their effectiveness and decide whether to standardise.

h. **Commissioning & Performance Review**

- Planned vs. actual commissioning activities.
- Post-handover performance issues and resolution plans.
- Seasonal testing outcomes.

i. **Commercial & Financial Review**

- Cash flow vs. planned S-curve.
- Accounts receivable and payable status.
- Contractual claims and dispute resolution.

j. **Organisational Learning & Asset Update**

- Lessons transferred to the central knowledge base.
- Templates, checklists, and trackers updated.
- Innovations documented for future reference.

k. **Project Resource Release**

- Formal release of project staff.
- Performance feedback documented in HR systems.

l. **Celebration & Closure Acknowledgement**

- Acknowledge project completion milestones.
- Recognize the team's hard work.
- Create a sense of accomplishment and closure.

Golden Rule: *"A closing meeting is not about celebrating the end; it is about preparing a stronger beginning for the next project."*

Sequence:

1. **Technical Closure** must come first → ensures the system is functionally complete.

2. **Contract Closure** follows → ensures all financial/commercial matters are resolved.

3. **Administrative Closure** runs in parallel → ensuring documentation, reporting, and organisational learning are not missed.

3. Snag List Management

No HVAC project closes without a **snag list** — a catalogue of unresolved issues, defects, or pending items that must be rectified before handover. Snags are normal in complex projects, but what differentiates a successful closure from a painful one is **how systematically these snags are identified, tracked, and resolved**.

Typical Snags in HVAC Projects

- **Ductwork**: Air leakages, missing supports, insulation not sealed, and access panels missing. Labelling not done

- **Piping**: Improper slope in drain piping, Drain leaks, welding defects, valve orientation errors.

- **Electrical**: Loose terminations, incomplete earthing, and missing identification tags. Final SLDs, missing Sign Boards

- **Controls/BMS**: Sensors not calibrated, VFDs bypassed, controllers not communicating.

- **Equipment**: Improper alignment of pumps, vibration issues in AHUs, cooling towers not balanced, supports missing for Fan coil units, Fan rotation wrong, Painting of pedestals, MS structures used

- **Site Readiness**: Ceiling closed before testing ducts, access not provided for maintenance. Closing all openings, making good

Example: In a hotel project, over 500 snags were listed, ranging from duct leaks to mislabelled cables. By holding daily snag review meetings with contractors and consultants, the team cleared 95% of items within three weeks and handed over the project on time.

4. Snag Closure Process

1. **Identification**: Snags logged during inspections, commissioning tests, or client walk-throughs.

2. **Categorisation**: Critical (must fix before handover) vs. non-critical (can be resolved during warranty).

3. **Assignment**: Each snag is assigned to a responsible subcontractor/team.

4. **Tracking**: Status regularly updated (Open → In Progress → Closed).

5. **Verification**: QA/QC engineer or consultant signs off on closure.

6. **Reporting**: Weekly snag reports reviewed in project meetings.

Practical Tip

Don't wait until the last week to create a snag list. Start preparing during the installation and commissioning phases — early identification saves time and avoids end-stage firefighting.

5. Sample Snag Tracker

Snag ID	Category	Description	Location	Responsible Team	Target Date	Status	Remarks/Verification
S-001	Ducting	Air leakage at flange joint	AHU-1 Supply Duct	Duct Subcontractor	10-Apr	Open	Smoke test pending
S-002	Piping	Chilled water pipe insulation incomplete	Basement Pump Room	Piping Contractor	12-Apr	In Progres	80% complete
S-003	Electrical	Earthing cable missing	Panel Room-2	Electrical Contractor	15-Apr	Closed	Verified by QA/QC
S-004	BMS	Temp sensor not calibrated	5th Floor AHU	Controls Vendor	14-Apr	Open	Awaiting spare

Figure 11.1 Sample Snags Tracker

Case Study – The Project That Never Truly Closed

A large commercial HVAC project for a **corporate office tower** was declared "completed" after the system started delivering cooling to all floors. However, beneath the surface, many closure steps were skipped or poorly managed.

- **Measurements Incomplete:**

  Ducting and insulation works were completed, but false ceilings were installed before joint measurements were taken. When the final bill was raised, the client disputed the quantities, claiming they were inflated. Since no proper measurement records existed, reconciliation became a battle of words.

- **No Amendments for Deviations:**

  During the project, additional AHUs and VFDs were installed on the client's request. The contractor executed the work but failed to raise variation orders or amendments in time. When final invoices included these items, the finance team at the client's side refused to honour them, citing a lack of approval.

- **Commissioning Incomplete:**

  The system was switched on, and "cool air" was demonstrated. However, seasonal testing and balancing were never carried out. Within months of occupancy, the client received multiple complaints of uneven cooling and high energy bills. Corrective work had to be carried out at additional cost, straining the contractor-client relationship further.

- **Final Payment Not Settled:**

  The client withheld the retention and part of the final payment, citing missing warranties, incomplete documentation, and unverified commissioning reports. The contractor, in turn, blamed the client for delays in approvals. The project dragged on in disputes long after occupants had moved in.

Key Lessons: ***Project closure is not the sole responsibility of the contractor.*** Owners and consultants also play a critical role in ensuring:

- Timely joint measurements.
- Approval of amendments and variation orders.
- Verification of commissioning and seasonal testing.

- Acceptance of complete O&M documentation.

> **Golden Rule:** *"Projects do not close themselves. Closure is a shared responsibility — when either side ignores it, both sides pay the price."*

11.3 Conclusion

Project closing is often underestimated in HVAC projects, yet it is the phase that leaves the **final and lasting impression** on the client. A system may be designed brilliantly and installed with precision, but if closing is rushed or incomplete, the project's value is quickly eroded by disputes, unpaid invoices, and poor system performance.

"**Project closure does not mean the project is forgotten**". The Defects Liability Period (DLP) or warranty phase ensures that the system continues to perform under real-world conditions. Smooth technical and contract closure makes it easier to transition into AMC (Annual Maintenance Contract) or extended O&M support. This continuity reinforces client trust and helps the contractor secure long-term business relationships.

When executed well, project closing is more than a contractual formality — it is a **professional discipline** that:

- Validates system performance against Owner's Project Requirements (OPR).
- Secures commercial and contractual settlements cleanly.
- Captures lessons learned and strengthens organisational knowledge.
- Builds trust, client satisfaction, and the foundation for repeat business.

> **Golden Rule:** *"A project is not truly complete when the cooling starts — it is complete when all parties confidently walk away, knowing performance, payments, and knowledge have been secured for the future."*

Table 11.3 Contract closing checklist

Category	Key Activity	Owner / Responsible	Status	Remarks
Technical Closure	Snag list prepared and closed	QA/QC Engineer	Closed	All 32 items resolved before handover
	TAB reports validated	Commissioning Lead	Closed	Verified by the consultant
	Seasonal testing completed	Commissioning Lead	Open	Scheduled for July
	O&M Manuals (as-built drawings, warranties, certificates) handed over	Documentation Lead	In Progress	80% compiled, awaiting VRF warranty docs
	Client O&M team training completed	PM/ Commissioning Lead	Closed	3-day training conducted
Contract Closure	Final BOQ reconciliation	Contracts Manager	Closed	All reconciliations signed off
	Variation orders approved	Contracts Manager	In Progress	Pending approval for 5 AHUs
	Final invoice submitted	Finance / PM	Closed	Submitted with the reconciliation sheet
	Retention release	Client Finance	Open	Held till warranty docs are submitted
	Subcontractor payments cleared	PM / Commercial	Closed	All 3 subcontractors paid
	Debris cleared, site demobilization done	Site Supervisor	Closed	Verified by client

Administrative Closure	Lessons learned documented	PMO	Closed	Added to org knowledge base
	Closing meeting conducted	Project Director	Closed	Minutes circulated
	Final project performance report prepared	PM/ Planning Engineer	Closed	Shared with management
	Resources released & performance feedback given	HR / PM	In Progress	2 engineers awaiting reassignment
	Templates/checklists updated for future use	PMO	Closed	New duct QA checklist added

Chapter 12: Future Trends of HVAC and Project Management

"The future belongs to those who prepare for it today. In HVAC project management, embracing innovation and sustainability is not optional—it's fundamental to mastering complexity and delivering excellence."

Learning Objectives

By the end of this chapter, readers will be able to:

- Understand emerging technology trends transforming HVAC systems, including IoT, AI, and digital twins.
- Identify new sustainability imperatives, such as decarbonization, energy efficiency, and indoor air quality standards.
- Apply advanced project management tools and techniques like AI-driven scheduling, BIM collaboration, and cloud-based PMIS.
- Recognise the importance of modular construction, prefabrication, and lifecycle commissioning in accelerating project delivery.
- Embrace digital transformation and data-driven decision-making for improved coordination, quality, and financial control.

- Anticipate future regulatory and cybersecurity challenges, incorporating them into proactive risk management.
- Develop the mindset and skills needed to lead in an increasingly complex, technology-enabled HVAC project environment.

12.1 Future Trends in HVAC Projects

Overview

The HVAC sector is rapidly evolving with advances in technology, rising sustainability demands, and changes in stakeholder expectations. The next decade will see a shift towards smarter, greener, and more digitalised HVAC solutions, particularly in India's fast-growing market.

12.2 Smart HVAC Systems and IoT Integration

- Systems are moving from basic automation to full IoT connectivity, where sensors measure temperature, humidity, air quality, and occupancy, feeding real-time data to integrated building management systems (iBMS)
- IoT-based solutions allow remote diagnostics, predictive maintenance, and adaptive controls, reducing downtime and operational costs, and optimising energy use.
- Engineers will need skills in programming, data analysis, and troubleshooting IoT devices to stay competitive.

12.3 AI & Predictive Maintenance

- Artificial Intelligence (AI) is used for predictive maintenance, identifying faults before they lead to breakdowns, and automatically scheduling repairs.
- Machine learning algorithms help analyse equipment performance data and detect abnormal patterns over time.
- Expect projects to demand integration between conventional HVAC controls and AI-powered platforms.

12.4 Energy Efficiency and Decarbonization

- Global and local energy codes—ASHRAE 90.1, LEED, WELL—are driving higher efficiency and lower operational carbon.
- Electrification, heat pumps, and hybrid systems are replacing gas-based models, while modern screw chillers and inverter-driven compressors reduce energy consumption.
- Engineers must consider lifecycle cost analysis, refrigerant selection (low-GWP options), and integration of renewables like solar, especially on large commercial sites.

12.5 Focus on Indoor Environmental Quality (IEQ)

- Post-pandemic standards prioritise IEQ, prompting adoption of advanced air purification (HEPA, UV-C sterilisation), demand-controlled ventilation, and real-time IAQ monitors in schools, hospitals, and offices.
- Projects increasingly require demonstrating compliance through sensor data and BMS integration.

12.6 BIM-Driven Collaboration and Digital Twins

- Building Information Modelling (BIM) is now essential for MEP design, enabling accurate 3D modelling, clash detection, and multidisciplinary coordination. appinventiv+1

- Digital twins—virtual replicas—allow simulation, remote diagnostics, and lifecycle performance analysis, improving stakeholder communication and facilitating rapid troubleshooting.

12.7 Modular and Prefabricated Systems

- Demand for speed and cost efficiency is driving modularisation and prefabrication of ductwork, piping, insulation, and air-handling units. This is already in maturing phase.

- Site-based installation is moving towards plug-and-play assembly, reducing risk, enhancing quality, and saving time. Reduce manual labor at site is the motto.

12.8 Advanced Commissioning and Analytics

- Commissioning now leverages wireless sensors and advanced analytics to verify equipment performance, automate defect logging, and ensure compliance with green standards

- The trend is towards continuous commissioning—monitoring and optimising building systems through their life, not just at handover.

12.9 Advanced Techniques in HVAC Project Management

12.9.1 Digital Transformation

- Project Management Information Systems (PMIS) integrate scheduling, documentation, cost tracking, and quality workflows, keeping all stakeholders aligned. marhy+1
- Cloud-based platforms enable remote collaboration, making project status visible in real-time across teams and geographies.

12.9.2 AI-Driven Scheduling and Resource Optimisation

- AI-powered project management tools forecast delays, optimise resources, and auto-generate alternative schedules based on equipment and workforce availability. marhy+1
- These systems help mitigate risks and improve accuracy in project delivery.

12.9.3 Augmented Reality (AR) & Virtual Reality (VR)

- AR and VR are revolutionising field work, enabling engineers and clients to walk through digital project models (Digital Twins), visualise intersections, and plan installations avoiding clashes before materials reach the site.
- These tools reduce errors, improve stakeholder buy-in, and cut down costly rework.

12.9.4 Integrated Lifecycle Costing and Value Engineering

- Sustainable project management now requires lifecycle cost analysis—factoring in operating expenses, energy, maintenance, and end-of-life impact, not just upfront capital.

- Engineers use energy modelling and scenario planning for decision making, bringing value engineering into every phase.

12.9.5 Regulatory Compliance and Green Building Certifications

- Modern HVAC projects must align with regional/global standards (LEED, BREEAM, WELL), requiring thorough documentation, test reporting, and ongoing monitoring.

- Project managers increasingly act as green certification champions, ensuring that design, construction, and commissioning align with sustainability goals.

12.9.6 Automation & Workflow Optimization

- Workflow automation (job cards, digital checklists, barcode-tracked inventory) streamlines task management and optimises repetitive processes—reducing the chance of error and freeing up site managers for strategic work.

- Automated dashboards and reporting tools improve clarity and accelerate decision-making.

12.9.7 Advanced Vendor and Subcontractor Management

- With modular systems and global sourcing, vendor performance tracking, digital delivery scheduling, and collaborative procurement platforms are becoming standard.

- Risk assessment and real-time vendor scorecards empower managers to act quickly on bottlenecks or subpar work.

12.9.8 Cybersecurity and Data Protection

- As HVAC systems and project management go digital, cybersecurity is a core requirement—protecting sensitive building operations and client data from intrusion.
- Engineers need basic awareness of secure data practices and compliance issues.

12.10 Actionable Recommendations by experts

- Invest in digital skills (BIM, IoT, AI, VR).
- Push for integrated, data-driven workflows that improve communication and transparency.
- Emphasise lifecycle sustainability and compliance as key project outcomes.
- Foster “soft skill” development for collaboration, negotiation, and stakeholder management in tech-dense environments.
- Stay abreast of regulatory changes to anticipate certification and market shifts.

12.11 Conclusion

The future of HVAC project management will be defined by digitalisation, sustainability, and integration. Advanced techniques—from IoT to BIM, AI scheduling, and automation—are not just trends but the new standard for delivering projects that excel in performance, efficiency, and client satisfaction. Project managers and engineers who master these tools—and align them with classic field discipline and team leadership—will be at the forefront of the industry's transformation.

Chapter 13: Summary & Conclusion

"Projects may finish on paper, but true success is measured in performance, trust, and reputation."

13.1 The Journey So Far

This book offers a hands-on framework for HVAC engineers to deliver successful projects by bridging the gap between theoretical project management and the day-to-day challenges of HVAC contracting. From design to delivery, every HVAC project is a story of interlinked decisions: contracts, planning, scheduling, procurement, installation, testing, and commissioning. This book aims to simplify that story into clear frameworks, tools, templates, and lessons that you can carry to your next site.

This book was never meant to be a theoretical manual. It was designed as a practical field guide for HVAC professionals — engineers, project managers, supervisors, and consultants — who face the daily challenge of turning drawings into comfort, efficiency, and reliability.

13.2 Key Takeaways by Section

Part I – Foundations

- HVAC project management is not optional — it's the difference between firefighting and foresight.
- The Project Manager is the conductor, aligning people, process, and performance.

- The HVAC lifecycle demands clarity at every stage, from idea to handover.

Part II – Planning Core

- Planning is the backbone: WBS, drawings, submittals, and method statements anchor execution.
- Scheduling links everything — civil readiness, procurement, commissioning milestones.
- Cost, risk, resource planning safeguard margins and timelines.
- Communication and integration are the glue holding all stakeholders together.

Part III – Execution & Control

- Execution = discipline in documentation + safe site readiness + synchronized procurement.
- Quality installation (air, water, plant, VRF, controls) comes only from process rigour.
- Monitoring and control are the steering wheel — EVM, dashboards, and issue logs keep the project on course.
- HSE is non-negotiable; safety lapses erase all other successes.

Part IV – Completion and Beyond

- Closing and handover transfer not just systems, but your reputation.
- Contracts and CCPM protect time, cost, and resources in today's environment.

- Lessons learned ensure today's mistakes don't repeat tomorrow.
- Critical Chain Project Management (CCPM) as a modern way to manage resource constraints.

Part V – Toolkits & Annexures

- Templates, trackers, checklists = ready-to-use weapons in your PM arsenal.
- Standards and case studies give you benchmarks and proof to work smarter.

Case studies highlight both successful and failed projects, stressing that every day spent in solid planning saves weeks in the field; conversely, shortcuts during planning invite delays, cost overruns, and disputes.

13.3 Project Execution: Where Plans Meet Reality

Execution is the most demanding phase, where even the best-laid plans face rapid-fire uncertainties.

- Documentation discipline → Only approved drawings/submittals at the site.
- **Site readiness & HSE:** Mobilise only when civil, safety, and utilities are in place.
- **Procurement sync**: Dispatch just-in-time, not before site readiness.
- **Quality installs:** Ducts, pipes, AHUs, VRFs follow method statements.
- **Cashflow clarity:** Joint measurements + timely billing protect margins.

- **Coordination**: Weekly multi-trade reviews = fewer clashes, faster progress.

13.4 Commissioning and Handing over

- Commissioning starts at planning, not at handover.
- Checklists + phased handover = predictable quality.
- Operator training + BMS integration = long-term performance.

Golden Rule: "A running plant ≠ a performing plant."

13.5 Lessons from Common Pitfalls

- Skipped planning means rework & overruns.
- Bulk/early procurement leads to wastage & blocked cash.
- Version chaos with drawings/submittals mismatch will lead to unwanted rework.
- Last-minute commissioning leads to inefficiency and customer complaints.
- If plant operator neglects and bypassed controls will lead to high energy bills and reduced plant life.

13.6 What Success Looks Like

Success in HVAC project management is not measured by whether the system starts on day one. It is measured by whether it continues to deliver comfort, efficiency, and reliability for years, while building trust with clients and strengthening your professional reputation.

- Meets scope, cost, time, quality → baseline success.
- Builds trust & relationships → repeat business.
- Leaves trained teams + documented systems → sustainable reputation.

13.7 Adopting Digital Tools and Modern Best Practices

- Scheduling & tracking → Excel (small), MS Project (mid), Primavera/ERP (large).
- Standard frameworks → **PMBOK + ISHRAE + ASHRAE**.
- Smart systems → **IoT + IBMS** = future-ready HVAC.
- Rule of thumb → Tools are only as good as the data & discipline feeding them.

13.8 Conclusion

This book closes, but your journey does not. Every project you undertake will present new challenges — new sites, new clients, new technologies. But the principles of project management remain universal: clarity, discipline, communication, and foresight.

Golden Rule: *"Do not just manage projects to finish them. Manage them to deliver value, earn trust, and build a reputation that lasts long after the project ends."*

Executive Recap:

Table 13.1 From Design to Delivery

Golden Rules	5 Lessons Learned	5 Pitfalls to Avoid
Start strong → Weak initiation = guaranteed trouble later.	Communication is the bloodstream of projects.	Treating commissioning as a last-day ritual.
Plan completely → Drawings, submittals, WBS, procurement drive success.	Procurement discipline protects cash flow & margins.	Rushing procurement without approvals.
Monitor continuously → What gets measured, gets controlled.	Integration of trades (civil, electrical, HVAC) avoids clashes & rework.	Relying on verbal/WhatsApp approvals.
Commission diligently → A "running" plant ≠, a "performing" plant.	Quality depends on processes as much as inspections.	Ignoring cost vs BOQ variances until too late.
Document everything → If it isn't written, it doesn't exist.	Reputation comes from long-term performance after handover.	Overlooking HSE — safety failures wipe out all else.

Takeaway: ***Projects will end, but your systems, professionalism, and reputation remain. Use this guide not just to deliver projects, but to deliver confidence, comfort, and credibility every time.***

Success Mantra: *Keep LEARNING. Keep reinventing yourself. There are no shortcuts to success. Be consistent in your efforts. Be open to adapting quickly!*

About the Author

RVSS Prasad, PhD

RVSS Prasad is a Life Coach and corporate trainer on Leadership and personal transformation. He has rich experience of over 35 years in Operations Management, Quality Management, Project Management, IT, Air conditioning and Supply Chain Management. He is a motivational speaker, Leadership Trainer, and executive coach.

He is a Mechanical Engineer with ME from **BITS Pilani**. He received his **PhD in management from JNTUH**, Hyderabad, for his work on improving the productivity of Software testing in select organisations using innovative techniques.

He is a visiting faculty at B-Schools of prominence, which include University of Hyderabad, JNTUH School of Management Studies, Symbiosis Institute of Business Management, NMIMS, VJIM, Hyderabad, and GITAM, Visakhapatnam.

His areas of expertise include:

- ✓ Project /Program Management
- ✓ Leadership Excellence
- ✓ Personal Transformation
- ✓ Emotional Intelligence

He has trained more than a thousand engineering professionals, IT professionals and mentored several resources in Leadership, Management, Project Management, and Airconditioning. He is a past president of ISHRAE Deccan Chapter and, past National Chair for Students at ISHRAE. He is currently working as a senior manager at Pavani Engineers, an Air conditioning Service provider.

Made in United States
Orlando, FL
05 November 2026

90230152R00246